OUR IMPROBABLE UNIVERSE

OUR IMPROBABLE UNIVERSE

A Physicist Considers How We Got Here . . .

Michael Mallary

THUNDER'S MOUTH PRESS
NEW YORK

Published in the United States by:
Thunder's Mouth Press
An imprint of Avalon Publishing Group
245 W. 17th St., 11th Floor
New York, NY 10011

AVALON

First printing June 2004.

Library of Congress Cataloging-in-Publication Data on file.

ISBN 1-56858-301-X
10 9 8 7 6 5 4 3 2 1

Typesetting by Pracharak Technologies
Printed in Canada on 100% recycled paper

Table of Contents

Introduction

Our Improbable Universe asserts that this universe is of inherent value whether or not it was created. The evolution of energy into matter, matter into life, life into mind, and mind into collective mind, is scientifically traced from the earliest moments of the Big Bang up to the present. This picture shows that a large number of physical phenomena had to be almost exactly as they are in order for us to have evolved out of the raw energy of the Big Bang. If the life-begetting substructure of our cosmos arose from a random process, then our fertile universe is a rare gem amongst an uncountable number of dead ones. If its physical laws were micro-engineered by a creator, the accomplishment is awe-inspiring. Either way, this incredible universe, and the life it has spawned, should be cherished. We are inherently part of an ongoing creative process that is majestic in and of itself.

Chapter 1
Fourteen Stepping-Stones

The energy of the Big Bang has spontaneously turned into people who can find joy in being. No inquisitive mind can ignore this fact. How did it happen? Why did it happen? How might the sequence of events have gone differently and failed to produce life? Is the universe the result of deliberate design? Was cosmic evolution a random chain of events? If so, what were the odds? Though it is impossible to provide complete answers, these questions cry out for our best effort using all we know today.

In the spirit of this curiosity, all human cultures have generated creation myths. These myths have provided insight into the origins and workings of the beautiful and capricious reality that has spawned our existence. Modern science comprising a vast body of observations and theories has greatly enhanced our understanding of our origins and fate. Through the scientific mind's eye, it is possible to see back into the history of the cosmos, and as we look in detail, we see an ever-lengthening list of characteristics that had to be just the way they were for intelligent life to have evolved. In other words, our extremely creative universe is also extremely improbable. *Our Improbable Universe* elucidates this assertion and explores some of its philosophical implications.

To understand the unlikelihood of our physical reality, with its biological riches, we have to follow the story of creation from the beginning to the present. About fourteen billion years ago, our universe exploded from a tiny point, a moment called the Big Bang. Though many of its details are still controversial, very few scientists dispute that the Big Bang happened. In the twentieth century, astronomers have established that all of the galaxies beyond our local cluster are rushing away from us. The most distant ones are moving away at nearly the speed of light. By mentally reversing this motion, one sees that all of the matter in this universe must have arisen from a single point in the distant past.

As scientific knowledge expands, consensus develops about the detailed history of the Big Bang. Currently the greatest controversy involves the details of what happened when the entire visible universe was smaller than a grapefruit, i.e. this is when the great-grandparents of matter were created out of the unleashed energy. This early matter later coalesced into the protons, neutrons and electrons that are the bricks and mortar of the atomic level of reality. The fact that any matter was left after only three minutes is actually one of the unsolved mysteries of science.

It is astounding that we live in a universe in which particles can spontaneously organize into people and other incredible beings. Adding to the wonder is the fact that this depends on an exact delicate physical structure that resulted from the Big Bang. Had this structure been slightly different in any of a large number of ways, the result would be a sterile universe. There would be no people, trees, or even bacteria.

In the last several hundred years, science has provided new ways to appreciate the beauty and complexity of existence. Unfortunately, a relatively small number of people have access to this way of seeing. Even practitioners of the scientific arts in many cases do not take a broad enough view to see the whole picture. Because this life-enabling substructure is not well understood by most people, it is often taken for granted. Life, taken for granted, is abused. The wanton destruction of our ecosystem is one example. The degradation of the human spirit in mass society is another. Possessed only with ignorance, we are as bulls in a china shop. Through knowledge of the universe, we can learn to love and respect humanity and our fragile biosphere.

In this book, I hope to provide a glance into this universe upon which life is based. This view will show that many balancing acts are taking place one on top of another to produce us and all that we see. The phenomena that led this universe to produce galaxies, stars, and planets are some of the stepping-stones on the path to our emergence. For example, just a slight change in the amount of energy and matter produced by the Big Bang would have resulted in a universe without galaxies or stars, let alone people.

In addition to the amount of mass-energy, there are at least thirteen other properties of our universe that were required to be the way they were for the Big Bang to have produced anything like us. One doesn't need to understand the science of these properties completely in order to appreciate how critical they are to our existence. Much of the science described here even baffles expert physicists. There are intricate dynamics beyond present human comprehension. The tip of the iceberg that modern science can see is incredible, but importantly, it hints at a deeper reality that is even more astounding. Here is a quick overview of the fourteen stepping-stones; more explanation of the importance of these phenomena is provided in later chapters.

Fourteen Stepping-Stones

1. Six Kinds of Quarks

The mass of your body is 99.95% nuclear matter, the clumped neutrons and protons of the nuclei of your atoms. Each particle is a clump of two different kinds of *quark* particles. In order for any of this mass to have emerged from the Big Bang, the universe had to be capable of producing at least six different kinds of quarks. Why? Because six quark varieties allows for a subtle asymmetry between the behavior of matter and antimatter. This lack of perfect symmetry is called charge conjugation-parity (CP) symmetry violation, put more simply as CP asymmetry.

2. CP Asymmetry and more

Though six kinds of quarks allow for a CP asymmetry, it does not require it. Without CP asymmetry, the Big Bang would have produced exactly the same amount of matter and antimatter. Then after several minutes, the matter and antimatter would annihilate each other leaving nothing but an expanding ball of light and the nearly inert particles called neutrinos. This would have been the sum total of the history of the universe. The future would contain nothing new after the first few minutes. But because of CP asymmetry, thanks for there being six quarks, for every billion particles of antimatter created in the Big Bang, a billion *plus one* particles of matter came into existence. All matter that we see today is that extra one part-per-billion left after an incomprehensibly huge conflagration of mutual annihilation.

But in addition to CP asymmetry the production of a surplus of matter depended on two more conditions. In the early universe there had to be an era when energy and matter were in a state of non-equilibrium and there had to be a mechanism for transmuting quarks and antielectrons into each other (see Chapter 2).

3. Just Enough Energy and Matter to Matter

When the part of our universe that we can now see was the size of a grapefruit, its density of mass and energy was just right relative to its rate of expansion: a difference of one part in a trillion trillion trillion trillion trillion would have precluded our existence. If the density had been higher, the Big Crunch, a contraction of everything, would have occurred too soon. The Big Bang would have been the Little Pop. Had the density been less than it was by the tiniest of fractions, the era of star formation would never have come to pass. Gravity would have been unable to overcome the outward rush, and the universe would have become a diffuse ball of isolated atoms, with no potential for congregating into stars. An explanation for this paradoxically precise amount of mass-energy density is one of the great triumphs of the *Inflationary Model* [1] of the Big Bang.

4. Just Enough Lumpiness

There had to be a slight amount of lumpiness in an otherwise smooth distribution of matter to form the seeds for galaxy formation. Some regions of space had to have slightly more matter than others. If the universe had been too smooth in relation to the expansion rate, then matter would not have pulled together to form stars. On the other hand, if the early universe had been too lumpy, then it would have become a very violent place. Gigantic black holes would have formed everywhere and gobbled up everything.

5. Four Forces

The complex behavior of matter is prescribed by four different forces, and the existence and relative strengths of these forces have been essential to creating a fertile universe. The most familiar is gravity, the effects of which are only visible on a scale much larger than our bodies. Importantly, it holds the Earth to its orbit around the sun. The raw energy of the Big Bang, from which all matter and energy emerged, came from the gravitational force. The coalescing of matter into planets and stars depends upon gravity.

The next most familiar force is electromagnetism. It makes magnets stick to refrigerators and lint stick to your clothes. At an atomic level, it causes electrons to orbit the nuclei of atoms. This force also allows atoms to stick together to make the thirty thousand complex proteins that our life depends upon. Our very thoughts are carried by neural impulses that are a complex mix of chemistry and electricity.

The last two forces are called "strong" and "weak." They are much less familiar to us because they operate only at subatomic distances. The strong force (also known as the nuclear force) glues protons and neutrons together to make complex nuclei. The weak force conspires with the strong to allow stars to burn hydrogen in a slow and steady way for billions of years. It changes protons into neutrons so that complex nuclei like carbon and oxygen can be built up. Without the weak force, giant old stars would not blow up and disgorge into space these important complex atoms, without which life would not stand a chance.

Life would be impossible without any one of these four forces. Each could be considered a stepping-stone to life in its own right. It is only a matter of bookkeeping convenience to group them as one very important stepping-stone.

6. Protons Don't Quite Stick

The existence of long-lived stable stars like the Sun depends on the relative strengths of all of the forces being just right. In particular, the nuclear force between two protons must be almost exactly what it is. If it were stronger by one-half percent, two protons would stick together permanently to form a helium2 nucleus [2] and throw off a lot of radiation. Fortunately, this does not quite happen. If it did, hydrogen would burn into helium at such a high rate that all stars would burn out in less a hundred million years. In this universe, helium2 nuclei hold together just long enough (i.e. 0.000000000000001 of a second) for our sun to burn hydrogen slowly and steadily, meaning it will go on for another four billion years. On the other hand, if the nuclear force were several percent weaker than it is, hydrogen burning would occur only in giant short-lived stars. In either case, there would not be enough time for anything complex to evolve. Though life might have been possible without stable stars, it would have been much less probable. If life's only stable habitats were chemical-rich hot springs, deep in the crust of the earth, then it is very doubtful that it could ever evolve from the primitive bacteria that live there into anything like us.

7. Helium Nuclei Don't Quite Stick

Two helium nuclei stick together for even less time than two protons. This is good. If they stuck more readily, helium would burn into carbon at such a high rate that stars would burn quickly and then blow themselves to smithereens. If they did not stick at all, helium would not burn into carbon and the other elements of life. Had this been the case, the only complex atoms remaining would have been the tiny residue of lithium from the age of nuclear synthesis that ended three minutes after the Big Bang.

8. Excited Carbon and Calm Oxygen

Carbon has an excited state of energy [3] that is just right to be reached when one more helium nuclei joins two others that are temporarily stuck on each other. This excitement in carbon allows it to hold together instead of breaking up into three helium nuclei most of the time. This enhances the rate of carbon production enough to stay ahead of its conversion into oxygen. Without this state at the right energy value, there would be very little carbon in the universe. The *absence* of a corresponding excited state in the oxygen nucleus is also fortunate in this regard. One could argue that life could evolve from a chemical system that did not include carbon, but carbon is extraordinarily suited to life's complex chemistry.

9. Big Stars Blow Up

About five billion years ago, in this region of the Milky Way, the vast majority of Earth's matter blew out of a giant star in a supernova explosion. As the song says "We are star dust." That stars explode the way they do at the end of their lives depends critically on many aspects of physics. These explosions spew into space complex elements that are cooked in the star's interiors, and these elements are the ingredients of the kind of rocky planet, with land, sea and sky, that we know and love.

10. Chunky Neutrons

Heavyweight neutrons are required for long-lived stars. If the mass of a neutron were less than that of a proton plus an electron, then the Big Bang would have produced a profusion of very heavy elements. There would be no lack of chemicals in that alternate universe. However, all hydrogen would be in a form that would burn rapidly in stars. Therefore all stars would be short-lived. The fact that neutrons are heavier depends on details of sub nuclear physics. If they were 0.1 % lighter than they are, there would be no long-lived stars in our universe.

11. Long Live the Proton

The most advanced theories of elementary particles predict that all protons and neutrons will eventually decay, making life impossible.

But fortunately, the decay of protons is so slow that it has not yet been observed. Nevertheless, we should not take long-lived matter for granted. Freeman Dyson has proposed that it will be the ultimate test of the adaptability of our species to cope with this eventuality. He speculates that our decedents might transform their material being from a dependence on atoms of ordinary matter to one based on electron-anti electron atoms. Whatever the likelihood of this wild idea, we have some time to get our act together. Experimental results indicate that this problem is at least ten billion trillion trillion years in the future.

12. 3D Reality

For life, three is the magic number where spatial dimensions are concerned. These are up/down, forward/backward, and left/right. To start with, three or less spatial dimensions are required to produce stable planetary orbits. If there were a fourth spatial dimension, the orbits of planets would not be stable in the manner of the Earth's orbit around the sun. In a universe that had four or more dimensions, orbits would be *metastable*. They would be like a pencil standing on its point. The slightest disturbance will make the pencil fall one way or the other. In a four-dimensional universe, the impact of a meteor on the dark side of a planet would cause it to gradually spiral in towards its star, where it would burn up. This is because the effect of gravity would grow in strength too rapidly, as distance decreased, for centrifugal force to prevent the planet's capture. Conversely, a meteor impact on the bright side of the planet would cause it to spiral out of its orbit. The strength of gravity would drop too rapidly to prevent escape into the dark and chill of deep space.

Theoretically, in a two-dimensional universe like the surface of this page, stable orbits can occur. However, complex neural networks, like our brains, would be impossible. The thousands of connections that each of our brain's neurons makes to each other would be blocked if the neural fibers (wires) could not cross over each other. Imagine a brain that consisted of neurons confined to the surface of this page. In order for a neural fiber to cross over another, to form a remote connection, it must lift off the page's surface. It must exploit a third dimension (up/down). The pulsations of a jellyfish-like creature would be its highest form of intellectual achievement. It is also unlikely that anything resembling the complex chemistry of life could occur in only two dimensions. The ability of carbon atoms to form bonds in multiple directions is crucial to many of life's processes. Therefore, two-dimensional universes are unlikely to be fertile either.

13. Wavy Matter

The wave nature of matter, described by quantum mechanics, is necessary for any universe that is more than a collection of black holes. Waves resist being squeezed into small volumes by forces and therefore resist collapse. The smaller the space that a quantum wave occupies, the higher the energy it has and the more it pushes out. This results in a pressure that resists compression. In the classical physics of particles (e.g. billiard balls), there is no such pressure. Classical matter would readily gravitationally collapse into black holes. In addition, waves are fundamental to the workings of matter and energy. None of the physical details described above would be anything like the same without this foundation. Our incredible reality is the interplay of countless shimmering waves.

14. Reclusive Particles

The tendency for atoms to form complex chemical bonds results from the Pauli Exclusion Principle, which states that two electrons cannot be in the same quantum state (e.g. the same place and veloc-ity) at the same time. Electrons, protons, and neutrons are loners and not groupies. (The exclusion principle depends in turn on the integrity of four other fundamental principles of physics that will be described later). In fact, all stable particles with mass obey the exclusion principle. Without this behavior there would be no com-plex chemical bonds, there would be no isolated particles as such. There would only be metals. Without exclusivity, groups of particles would congregate in ultra-dense gravitationally bonded clusters. As such, they would easily collapse into black holes and that would be that.

The above recipe of fourteen essential ingredients for a universe in which sentient beings can evolve is undoubtedly incomplete. As science advances, the list will lengthen. Here are some other condi-tions without which life would have been hard-pressed to emerge, if it could have been possible at all.

Long-lived Natural Radioactivity

Long-lived radioactive elements like potassium40, uranium, and tho-rium are needed for the renewal of carbon dioxide in our atmosphere in the distant future. The decay of these elements in the earth keeps volcanoes active by replenishing the primordial heat of the earth. Volcanoes in turn replenish carbon dioxide in the atmosphere, an element without which carbon-based life would be in big trouble.

Vast Stellar Spacings

We should thank the stars for keeping their distance, for if stars were more closely packed, near collisions would be more frequent. These close encounters would destroy planetary systems by fling-ing the planets into deep space or into eccentric orbits with highly variable climates. This happens in the densely packed centers of galaxies and in globular clusters of stars. Fortunately for us, it does not happen often in our outer galactic region.

Ice Floats

Most solids sink in a pool of their liquid, but oddly ice floats in water. This is very good for water-based life. Chilly oceans such as on Europa (a moon of Jupiter) develop a thick crust of ice that insulates them from the deep freeze of space. Under the ice, life stands a chance. Recent studies indicate that on Earth global freeze-overs have happened many times [4]. If water froze from the bottom up instead of from the top down, life here would be a great deal more primitive than it is. Only creatures that could ride out being frozen solid would survive.

This list of nonessential ingredients could go on much longer. Many are specific to evolution on Earth [5]. But the list of fourteen essentials already demonstrates that this universe is astonishingly suitable for life. Truly, the uncanny coincidence of all these factors seems to give the appearance of deliberate design. In the nineteenth century, William Paley saw the skill of "the Master Watchmaker" in all of the wonders of nature that were apparent then. Today we know more of the tightly organized mechanics of the universe, and have ever more to wonder at.

But a deistic hypothesis is not the only explanation for a fine-tuned universe that has produced life. There are many scientists who propose that this beneficent structure has been determined by a random unknowing process. While a fertile universe is improbable, there might be a huge number of universes, each with its own physical laws, being produced by a random process making the emergence of one universe with all "the right stuff" more likely. Prior to current inflationary theories of the Big Bang, the Nobel laureate John Wheeler proposed that the universe may pass through many cycles of Big Bang explosions followed by Big Crunch contractions. In each cycle, the newborn universe might have an entirely new physical substructure. After a trillion trillion cycles a fertile universe with sentient beings could result.

Inflationary models hypothesize that a large number of universes are being spontaneously created *in parallel* within a gigantic metauniverse. The metauniverse is like a huge frothing sea with in which each universe is like a bubble of steam rising and expanding from the bottom of a hot pan of boiling water. Just as the Milky Way galaxy is one galaxy in our observable bubble of trillions of galaxies, our observable universe may be only one in a larger metauniverse that contains trillions of universes. They may be like our own or different from it in any or all of the fourteen respects discussed above. If they are different, the vast majority are probably also sterile. In both (cyclic or parallel) approaches it is possible to imagine that at least one universe in a trillion trillion had just the right properties for life to evolve. In a universe that produced observers, they would marvel at the apparent providence of its physics. It would seem to have been designed, even if it hadn't been. This particular interpretation of the apparent compositional wholeness or design of the universe is known as the *anthropic principle*. Any universe that produces witnesses would appear to be designed, in part because the botched universes are not observable to the inhabitants of the successful ones. We can view them only in the mind's eye.

A statistically-based hypothesis for generating fertile universes brings to mind the old monkeys and Shakespeare game. Imagine a huge number of monkeys banging away at typewriters. The goal is for them to generate a work of Shakespeare through sheer random persistence. In a trillion trillion attempts they can get "to be or not to be." If each attempt requires three monkey minutes (the time it took the Big Bang to create the nuclei of some simple atoms), and there are a billion monkeys, then it will take a billion years to get this phrase. If only this successful attempt had an audience, then the huge number of failures would only matter to the monkeys (and the monkey managers who had to keep them at it for a billion years). In this hypothesis, the monkeys are a mindless random process that is not concerned with speed and efficiency. Yet when all is said and done, an observer might still find "to be or not to be" in the babble and declare it to be the work of God.

So here is where the battle line is drawn in the debate over whether or not the universe was created. If a phenomenon is perceived to be fine tuned for life at the level of one part in a hundred, then a proponent of the anthropic principle simply proposes a hundred times more randomly generated universes to make the additional requirement statistically feasible. The compound odds against getting a life-friendly universe keeps mounting. But there is no shortage of bubble-generating points of instability in the metauniverse like the one from which our universe popped into existence. Alternately, in a pulsating universe, it is merely a matter of waiting out a hundred times more cycles of expansion and collapse before a "hospitable" universe arises. If there are no witnesses during the sterile cycles, then the process wouldn't seem too long. Even a creator, with enough patience, might choose, or have no choice but, to create in this fashion.

It would appear to be impossible to prove the presence of the hand of the Master Watchmaker from the intricacy of the watch. However, one might hope to find evidence for the artist's signature on the canvas. Or perhaps a built-in bias that reflected a designer's taste, rather than a functional part, might hint at an unseen hand, some quirk of nature not required to produce life and that was inherently improbable in a universe that was spawned by a random process.

Though many have searched for such a signature, nothing conclusive has been found. Perhaps, a hint of a "designed bias" comes from the existence of uranium. It did not have to exist in a universe that evolves intelligence. Did a creator choose to include it in our universe for a purpose? Is the potential for nuclear holocaust in our universe a kind of cultural filter? Is it here to allow incorrigible barbarians to eliminate themselves? Is it a teacher? In fact it already has been a teacher. Because of the many harrowing experiences provided by forty years of the Cold War, we have learned a few things. In particular, we now know that the time-honored ethos of parasitic empire-based domination can't make it in the long run in this universe. Does this indicate that a creator had second thoughts about the kind of creature that would result from unbridled natural selection?

On the other hand, the existence of uranium in our universe might be seen as an unfortunate quirk of chance resulting from a random and mindless universe begetting process. This search for a signature begins to look and feel like a cat chasing its tail. The issue is unresolvable in a scientific sense. One can only choose a course and follow it through to its conclusions. So let's choose both courses one at a time and see where they go.

If this universe is the result of careful planning, then the creator went to a lot of trouble to fashion it in just such a way that particles like protons could turn into self-aware beings. The creator also built in enough time for this to happen. In this deistic view, life is a precious yet seemingly mixed blessing (death comes with it). We must therefore make the most of it. Clearly we should seek harmony with the ongoing process of creation by participating as fully and creatively as we can. With the destiny of life on Earth in our hands, we have a deep responsibility.

On the other hand, even if this universe is the result of a cosmic throw of the dice, and not deliberate creation, then we have won an incredibly improbable jackpot, a universe that has the potential for almost limitless creativity. Clearly we should not squander our winnings. Instead, we should seek to become part of this beautiful and rare creative process to the best of our ability, encouraging the growth of richness around and within ourselves. If we blow it by destroying the Earth, we will have changed the last digit of our winning lottery number from a winner to one of the infinite number of losers both for ourselves and for the other species with which we share this remarkably fecund planet. Whether a creator is behind existence or not should be irrelevant to human behavior. In either case, we should cherish the improbability of being here. We should integrate our lives with the ongoing creativity that surrounds us. We should try to become a constructive part of the closest thing to eternity that we know we have. And we should marvel at the ongoing miracle of it all.

Chapter 2
Bang!!

All that we see exploded from a ball the size of grapefruit at nearly the speed of light about fourteen billion years ago. The temperature was just shy of a million trillion trillion degrees. The density of particles was a trillion trillion trillion trillion trillion times that of your body. We can infer these things using Einstein's Theory of General Relativity combined with evidence from gigantic telescopes and particle accelerators. Telescopes tell us that galaxies in deep space are rushing away from us at more than 95% of the speed of light. Therefore, they must have blown out of a super-dense region long ago.

Scientific efforts to see back to the grapefruit-sized ancestor of the visible universe are similar to prehistoric archeology. A great deal can be discerned, but any conclusions we draw become progressively more sketchy as we go further into the past. Nevertheless much of, the history of the universe can be interpolated from what we can see today. The issue that challenges our confidence most, however, is the paradoxical *flatness* of our universe. Flatness means that lines that start off as parallel will remain so to infinite distance. This is not true of the longitudinal lines of the globe (they meet at the poles) because it is not flat. Similarly, the universe as a whole could have been *curved* (in a *closed* way) and with a very powerful telescope (and a lot of time) you could see the back of your own head.

More importantly, the flatness we see implies that the expansion rate and the mass-energy density are well balanced, which implies that they were extremely well balanced when the grapefruit was ripening (its growth spurt was over), so to speak. Had that incredible degree of balance (i.e. one part in 10^{60}) not been achieved at that time, we would not have been given the chance to exist. As described previously, either a premature big crunch or a star-forbidding excessive expansion rate would have resulted.

Microwave images show a *homogeneity* (i.e. it looks the same in every direction and at every distance) of the universe which is also paradoxical because light could have traversed only a trillionth of a trillionth of the grapefruit's diameter during its short life. How could all parts of it be so similar if they could not be mixed together (by light)? Though it is very homogeneous or smooth, there is some lumpiness (i.e. density fluctuations). This lumpiness is just right for the begetting of life. The density fluctuations are at the level of fifty part per million. If they had been much greater we would have had a life-destroying super abundance of black holes. If these lumps were much less there would be very few stars. What kind of cosmic Mix-Master could achieve this just right lumpiness?

Possible answers to these two paradoxes have followed from theoretical attempts to unify the electromagnetic force, the weak force, and the strong force into a single force that would act during the "grapefruit era." But proving any of these theories, known as Grand Unified Theories (or GUTs), has so far been an elusive goal: we are far from replicating the temperatures of that era—even a particle accelerator seventeen miles in circumference will achieve (in 2007) temperatures that are only a trillionth as hot.

All GUTs, share the assumption that the energy density of space was immense at the start of the Big Bang. When Alan Guth combined this postulate with Einstein's theory of general relativity, he discovered that gravity was a repulsive force during the earliest moments of the Big Bang. This led to an era of runaway inflation during which the size of the universe doubled at least a hundred times. During this inflationary era, the first 10^{-35} seconds, the vacuum energy that drove the inflationary process rapidly transformed into the hot plasma of particles that classic Big Bang cosmology deals with. The transformation was complete when all that we survey could be enclosed by our hands. This sequence of events during this brief time is illustrated in Figure 2.1 below.

We have at least as many theories of inflation as GUTs. However, it is a great achievement that these inflationary theories account for the flatness and homogeneity of our universe naturally. These theories hypothesize an initial energy-packed medium, the *false vacuum*, that was (and still is) inflating because gravity is turned into a repulsive force by this kind of vacuum energy (hypothesized by GUTs). The tiny region that inflated into the grapefruit was small enough for light to propagate across it. This accounts for the homogeneity of our universe. The extreme expansion of space automatically flatted out any curvature that existed at the birth of things. If our primordial universe was curved, inflation blew out the distance at which parallel lines obviously converge or diverse to much larger than the visible universe. It would be as if the world were so large that Magellan could never circumnavigate it in order to prove that it was a sphere. An explorer going north thousands of miles would experience no climate change, north and south poles being billions of miles apart. Similarly, lines of longitude would appear to be parallel instead of convergent at the poles.

In an absolutely "flat" universe, the expansion rate gradually slows towards zero, but never quite stops. It is like throwing a ball straight

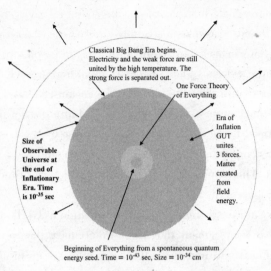

Figure 2.1 The sequence of events in an inflationary universe begins with the Plank Time and proceeds up past inflation to the Classical Big Bang. After this time the established theories of modern physics are applicable. Before this time we are still guessing. During the GUT phase of inflation, energy and matter transformed back and forth in a process that cooked up the particles upon which our material reality is based. As the ball of energy expanded, the GUT force fragmented into the strong force and the electroweak force. The energy released, cooked up even more matter and super heated it. At this point the inflationary era was over. The universe that is visible to us had expanded to about the size of grapefruit in less than a billionth of a trillionth of a trillionth of a second. From then on, the uiverse expanded according to conventional Big Bang dynamics.

up with just enough speed to go forever (Earth's escape velocity is 25,000 mph). At the end of the inflationary era, the expansion rate was precisely right for longevity, i.e. if it had been a tiny bit lower (by 1/1,00...00 with 60 zeros) the expansion would have reversed long ago and Earth would have been annihilated in the Big Crunch before humans had evolved. If it had been higher by a comparable amount, hydrogen atoms would have moved away from each other too rapidly to gravitationally coalesce into stars. Within observational errors, our universe is a very "flat" place and we like it like that.

At the end of the inflationary era, variations in density existed that would later spawn super clusters of galaxies, galaxies, and stars. These variations were slight lumps in the otherwise smooth distribution of matter and energy. They must have existed, because we exist. But in order to gain understanding of this, we must go back further in time, before the era of inflation (labeled as the TOE era in Figure 2.1). Understanding what happens here requires a Theory of Everything, or TOE.

The abundance of GUTs in our time is well matched by the number of TOEs, which deal with a temperature that is a thousand times higher and thus is even less experimentally accessible. At such high energies and small distances, the force of gravity becomes comparable to the GUT force. Therefore all forces are united into one force in the TOE era. In addition to achieving the unity of all forces, TOEs also accomplish a consistent merger between Einstein's very successful Theory of General Relativity and the even more stringently tested theory of quantum mechanics. In our reality gravity is observable action on very large objects, and quantum mechanics on very small ones. At small dimensions the wavelike nature of matter and energy becomes apparent. During the TOE era, the quantum nature of reality produced wavelike fluctuations in the energy density of space that were the seeds for the formation of galaxies and stars. Inflation smoothed out these fluctuations enough so that the universe did not become dominated by super violent black holes. Without the conspiring of quantum fluctuations and inflation the universe would not have been nearly as hospitable.

The smoothness or homogeneity of the universe has been observed in the positions of over a million galaxies as mapped with gigantic telescopes. On a large scale, the distribution of these galaxies in deep space is quite uniform (homogeneous). It is like a milkshake that has just a little bit of lumpiness left from the ice cream. Amazingly, a region once so microscopic that it was only a hundred or so quantum wavelengths wide is now the size of the visible universe and each microscopic wave crest has become a supercluster of galaxies 500 million light years across.

The just-the-right-lumpiness of the early universe was first observed in the microwave background emission of deep space by the Cosmic Orbital Background Explorer (COBE) satellite. Variation in microwave intensities, at the level of parts per hundred thousand, reveal the corresponding temperature and density variations of a time three hundred thousand years after the Big Bang. At that time the temperature had fallen below 3000° Centigrade and all of the free electrons in the plasma had attached to atoms. Soon these atoms stopped emitting light and became transparent. Since then, the universe has expanded a thousand fold. Expansion cooled the hot photons flying around into the cold microwave background radiation (−454F) that we observe today. George Smout, who headed a team of COBE scientists, said: "It is like seeing the face of God." This got him into hot water—some colleagues disagreed with his interpretation or felt that his religion should not be mixed up with science. But the awe that he expressed was appropriate whether the universe was a divine creation or whether we just happen to live in a universe with the right kind of inflation.

In the inflationary models, the vacuum that generated our universe is unstable. The process that got ours going is happening all over the place in the same way that a pan of boiling water has many bubbles of steam popping into being all the time. What appears to us as a Big Bang is one of many relatively little pops in a larger metauniverse. This larger universe is forever expanding and forever blowing bubbles. It is like a froth of bubbles in a gigantic wave that is crashing about in a sea of unimaginable proportions. Our universe is just one bubble and we can see only a tiny fraction of that bubble. There is no hope of ever seeing other bubbles. These other universes, with physical laws that may be different from ours, are carried away from ours by the continuing process of inflation of the metauniverse. This is illustrated in Figure 2.2 below.

The bubbles in the metauniverse may be initiated by the same kind of quantum fluctuations that lead to the slight lumpiness that was essential to star formation. But in this case, the quantum perturbations, in an unstable system, spawn whole universes with unique physical laws. For every bubble universe that possesses laws that allow life to evolve, there are an uncountable number of barren ones with no stories to tell.

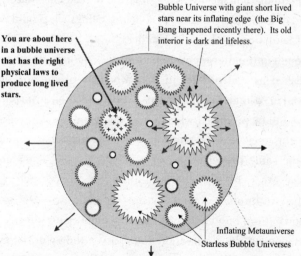

Figure 2.2 The forever inflating metauniverse which contains many bubble universes such as our own. Inflation has ceased in the interiors of the bubbles so the laws of nature have settled into a stable state there. However inflation continues at their boundaries where the GUT field energy is still converting into a plasma of hot particles as the bubble grows. Quantum effects rough up the edges thus producing the inhomogeneities that spawn the formation of galaxies and stars.

Some have speculated that this larger metauniverse may have been spawned by quantum fluctuations in an even more astounding space with at least nine spatial dimensions (directions), and one that corresponds to our time. These extra dimensions are needed to unite Einstein's well established theory of General Relativity with the more recent theory of quantum mechanics. One idea is that during the inflationary era the three dimensions of our bubble universe, inflated to the size of a grapefruit while the extra dimensions remained *compacted* at a miniscule undetectable size. But don't expect to understand it. No one does. I described it briefly for you to appreciate how complex the foundations of reality might have been just to get a universe to a ripe old age of a millionth of a second. Words like mysterious and mystical come to mind even though there are equations that speculatively describe it. The fact that a minute portion of it would eventually evolve into beings who could discover these foundations adds to the wonder.

Though antimatter seems to be a science fiction invention, it was established as real physics in 1932 when its production was first observed in cosmic ray interactions. Antielectron-electron pairs were observed to be produced when a very energetic X-ray (gamma ray) struck the nucleus of an atom. It was discovered that antielectrons have the same properties as electrons (e.g. mass, spin, magnetic moment) except for their charge; antielectrons are positive instead of negative. Antiprotons are negative instead of positive. Due to cosmic rays and natural radioactivity, there is antimatter in your house right at this moment. Its observation confirmed a theory of the electron, proposed by Paul Dirac in 1928, which predicted antielectrons. The Dirac theory also predicted that electrons cannot be made without making an equal number of antielectrons. The Dirac theory was later refined into the theory of Quantum Electrodynamics, and has been tested to greater precision than any other theory that we have today. No variation from its predictions has ever been observed.

In all experiments to date, exactly equal amounts of matter and antimatter have always been produced in energetic collisions of particles. As the antimatter moves around in the apparatus, it inevitably latches on to some matter. In an instant the two particles annihilate each other and a pulse of energy is released typically as light. In the end there is never any net excess of matter produced. This is illustrated in Figure 2.3.

If particles and antiparticles are always produced in pairs, how did the Big Bang produce any matter at all? Why didn't all the antimatter annihilate all the matter in the first few minutes? Why isn't our universe a lifeless ball of massless particles like photons (light),

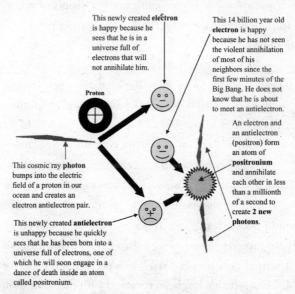

This newly created **electron** is happy because he sees that he is in a universe full of electrons that will not annihilate him.

This 14 billion year old **electron** is happy because he has not seen the violent annihilation of most of his neighbors since the first few minutes of the Big Bang. He does not know that he is about to meet an antielectron.

Proton

An electron and an antielectron (positron) form an atom of **positronium** and annihilate each other in less than a millionth of a second to create **2 new photons**.

This cosmic ray **photon** bumps into the electric field of a proton in our ocean and creates an electron antielectron pair.

This newly created **antielectron** is unhappy because he quickly sees that he has been born into a universe full of electrons, one of which he will soon engage in a dance of death inside an atom called positronium.

Figure 2.3 The production of matter and antimatter from light (e.g. a cosmic ray photon) followed by its annihilation. This is shown as happening in the ocean but it has also occured many times in your home while you are reading this.

expanding to infinity forever? This is a mystery that has been slowly unraveling for four decades and still has a great distance to go. Final proof of any theory will be very difficult, in part because the energy level (temperature) of the reactions that might produce an excess of matter is a trillion times higher than people can presently produce, for such was the temperature of our improbable grapefruit at the end of the inflationary era.

We probably will never create a device generating more matter than antimatter, but we can still get indirect insight into this issue. People knew of the existence of atoms for a century before they could observe their shape with the Scanning Tunneling Microscope a decade ago. Some suspected their existence thousands of years ago. The mind's eye is the most powerful microscope. By piecing together disparate evidence we can create a vision of what might be. This human ability is one of the most inspiring products of this incredible universe.

As mentioned before, the production of surplus matter depended on CP Asymmetry. Its story goes back to the early 1950's when nuclear scientists believed that nature was bilaterally symmetric like the human body. If you wanted to know what the left half looked like you only had to look at the right half through a mirror. For example, "MM" has positive mirror symmetry. It looks the same in a mirror. This is known as positive Parity Symmetry (P). The decay of neutrons into protons dashed the hopes of nuclear theorists that nature would be so simple, for this decay process was observed to be clearly asymmetric when viewed through a mirror. Instead it has negative parity. For example "MW" viewed through a mirror looks like "WM." The left side has been reversed with the right. It appears to be turned upside down relative to the right side. This turned the world of nuclear theory upside down for awhile.

Tranquility was temporarily restored when it was noted that if the electric charges (C) of the particles were reversed (plus to minus and minus to plus), then the reaction *was* mirror symmetric. The weak force appeared to respect CP Symmetry. The life of the theorists was simplified until an experiment by Val Fitch and James Cronin in 1964 dashed their hopes once again. They later received the Nobel prize for their observation in which CP Symmetry was violated at the level of parts per million. Though the lives of theorist seemed less bearable, everyone's life became possible. If CP Symmetry could not be violated then matter and antimatter production would always be exactly equal. Then we would not be here to be perplexed by the issue. But this difference, this slight dominance of matter over antimatter was like revealing a crack that was just big enough for a material universe to have poured through [1].

The exact source of the CP Asymmetry is still a mystery. However the reason that it can exist at all, in subnuclear physics, has become more apparent after experimental and theoretical developments of the mid 1970's. These developments are related to the Quark Theory of matter [2], which states that protons and neutrons, the building blocks of the atomic nucleus, are themselves made out of particles called quarks. Two different kinds of quarks go into making protons and neutrons. They are called Up and Down quarks. A proton has two Ups and a Down. Neutrons have two Downs and an Up. The Ups have +2/3 of an electric charge unit (the electron charge) while the Downs have −1/3 of a charge. Therefore the proton comes out with +1 for its charge (2/3 +2/3 −1/3 = 1) and the neutron has no charge (2/3 −1/3 −1/3 = 0). This is illustrated in Figure 2.4 below.

At the time that this Nobel-Prize winning theory was proposed, there was also a third quark called Strange. It was postulated in order to explain a class of particles that were known as Strange particles. They got this name because they were not found in nature except in strange looking cosmic ray reactions. This amusing terminology unleashed a flood of whimsical names for the bizarre beasts that subsequently appeared in the particle physics zoo (e.g. Charm, Truth, Beauty, Red, Blue, Green, Wimps, etc.).

Since that time, three more kinds of quarks have been proposed to account for additional classes of observed particles. The corresponding new quarks are called Charm, Bottom (also called Beauty), and Top (or alternately Truth, to go with Beauty, as in John Keats' "Truth is Beauty and Beauty is Truth"). Evidence for the Top quark was found only recently, more than twenty years after the Bottom quark was found. Because the quarks fit into pairs, the existence of this sixth quark was a forgone conclusion once the Bottom quark was found. Six quarks were built into all of the GUTs from the start because they were part of the *Standard Model* of quarks that is the present frame work of particle theory.

One very convincing piece of evidence for the sixth quark is that you and I are here today. Without at least six different kinds of quarks, the CP Asymmetry in quark reactions cannot occur. The theory of this connection was proposed by M. Kobavashi and J. Maskawa in 1973 at a time when only three quarks had been observed. In this theory, the existence of six kinds of quarks makes it possible for a CP Asymmetry to exist. Thus, the existence of at least six kinds of quarks is the stepping-stone which sits beside the stepping-stone of CP Asymmetry.

Though the existence of six quarks allows the CP asymmetry it does not require it. Similarly, though CP asymmetry allows a surplus of matter it does not require it. Nature must meet two other conditions for this to happen. The second requirement is that inflation must pass through a period of thermal dynamically non-equilibrium [1]. Energy must be preferentially transferring from one

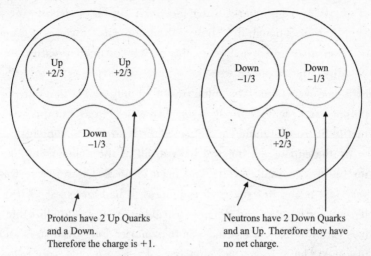

Protons have 2 Up Quarks
and a Down.
Therefore the charge is +1.

Neutrons have 2 Down Quarks
and an Up. Therefore they have
no net charge.

Figure 2.4 The Quark Theory proposes that nucleons are made of three quarks. Protons (q = +1) and neutrons (q = 0) are made out of Up quarks (q = +2/3) and Down quarks (q = –1/3). Unstable nucleons can also contain quarks that are called Strange (q = –1/3), Charm (q = +2/3), Bottom (q = –1/3, also called Beauty), and Top (q = +2/3, also called Truth). Colored Quark Theory proposes that each quark in a nucleon had to be colored differently from the other two with either Red, Blue, or Green.

state to another. GUT theories and the transitory nature of inflation provide this. The energy in the GUT fields is converted into particles and light. The non-equilibrium between the GUT field energy, light, matter and antimatter makes the production of a surplus of matter a one way street. Without this non-equilibrium the energy would slosh back and forth between these four categories until an equilibrium was established in which there was no surplus of matter. When all of the field energy is used up inflation ceases because there is no more vacuum energy with which to make gravity repulsive.

The final requirement for getting a surplus of matter is that nature possess the ability to transmute quarks into the electron like particles called *leptons* (i.e. electrons, muons, and taus). The surplus of quarks over antiquarks is matched by a surplus of electrons over antielectrons. Though this ability is built into GUT theories it is not a universal requirement of any theory. But fortunately nature fulfills this requirement as well so we have our surplus. However, the existence of this quark to lepton transmutation implies that our surplus of quarks will eventually decay into antielectrons, but this eventuality (if true) is unimaginably far in the future (see Chapter 3).

By the end of the inflationary era, all of the major conditions for life to eventually evolve had been created. The universe was very flat, it had just the right lumpiness, it had an excess of matter over antimatter, and it had physical laws that would allow life's necessities to evolve (e.g. like complex matter, long-lived stars and planets). The universe had already cooled to a temperature below which the excess of matter over antimatter could form. Matter at this stage consisted of a sea of the six quarks and leptons, along with their antiparticles [2]. By about a hundredth of a second after the Big Bang, the temperature and density had fallen to where quarks condensed into clumps of three of the "familiar" quarks. These three-quark clumps were neutrons and protons (Figure 2.4). This was the raw material for the nuclei of all our atoms still in existence today. There were also analogous nucleons made from the four unfamiliar quarks. These nucleons were unstable and quickly converted into neutrons and protons when the temperature became too low to regenerate them.

As the temperature fell further, the renewal of antiproton/proton pairs in high energy collisions ceased and the one part per billion excess matter became the only nuclear matter in the universe. At that time, normal matter consisted of free neutrons, protons and electrons plus some antielectrons. These were still being made in electron/anti-electron pair production that resulted from the collision of high energy light with electrons and protons (see Fig. 2.3). After fourteen seconds the temperature had fallen past the point (a billion degrees centigrade) where electron/antielectron pairs could be produced. At this time free neutrons and protons began to stick together to form simple nuclei. Various forms of hydrogen and helium nuclei were produced in abundance, along with a tiny amount of lithium. Essentially none of the elements of life such as carbon, oxygen and nitrogen, even existed. All of the chemical elements of life would be produced about two hundred million years later, when stars began to form.

After about three minutes, the temperature of the universe had fallen below ten million degrees Centigrade. At this cool temperature, protons could no longer join to form heavier nuclei. The universe continued expanding for another several hundred thousand years before its next major change in character occurred. By this time the temperature had fallen below several thousand degrees. The free electrons and nuclei, which had formed the opaque plasma began to stick permanently to each other to form electrically neutral atoms. The sudden condensation of the plasma into atoms made the universe transparent to the propagation of light (neutral atoms do not scatter light readily). For every atom that existed at that time, there were a billion photons of light, and this ratio holds true today. The COBE satellite's count of microwave photons yields this number. Interestingly this is the same ratio as that of surplus matter, compared to antimatter, created during the inflationary era. These primordial photons are still propagating around the universe today. However, because of the great amount of expansion that has occurred in the intervening fourteen billion years, these photons have been cooled from a temperature of 3000° Kelvin (6000°F), down to *three* degrees Kelvin (–454F). They have "red shifted" right out of the visible spectrum and can now only be detected with microwave radio receivers. They were first detected by A. Penzias and R. Wilson in 1965 as an excess of microwave radio noise in space. Penzias and Wilson were trying to establish the first microwave satellite communications link at that time. These microwaves from deep space (they did not know it at the time) were high on their list of stubborn sources of extra microwave noise that they had to eliminate in order for their receiver to meet theoretical sensitivity expectations. After they evicted a family of pigeons from their horn antenna, the errant signal was still there. When they consulted with the Princeton cosmologist Robert Dicke, he pointed out that Big Bang cosmologists expected this signal even without a nest of pigeons. He knew that this noise might be the cooled-down photons from the age of condensation. This observation actually was the basis for wider acceptance of the Big Bang theory of the universe. The "lumpiness" detected by the COBE satellite has been further validation of the theory.

So far we have discussed five of the stepping-stones to an intelligent universe [3]. These were all involved with setting the stage for the era in which star formation could begin a billion years after the very beginning. The Big Bang went through many sequences of crucial events even within the very brief inflationary era: the four forces each acquired their own identity; the universe brewed up just the right amount of matter out raw energy; the formation of nuclear matter resulted from the nuclear force; atoms formed several hundred thousand years later due to the electric force; and finally, the gravitational force formed stars and solar systems a billion years later. The first generation of stars had no complex atoms or complex planets and were therefore devoid of life. Yet another sequence of events would generate the complex reality that we know as the chemical basis of life.

Chapter 3
Strings and Quantum Things

Before we move on to a description of stars, let's consider further stepping-stones that allowed life as we know it to be possible in our universe. Some of these life-giving phenomena are so basic that it is hard to see how a metauniverse could do without them. One of the most fundamental is the wavey nature of the microscopic world upon which our reality is built. Modern physics has determined that waves are at the heart of all physical reality. What we perceive to be a hard table is in fact a shimmering collection of ten thousand trillion trillion waves colliding with each other. Their collective dance locks them into the shape of a table. The study of the wave nature of reality is called quantum mechanics. In quantum mechanics waves are everything. Massive particles are in fact waves that are trapped in forms made of complex interactions of other waves. These interactions are defined by forces that have a wave nature as well.

The theory of quantum mechanics is difficult to comprehend. This is because we are used to thinking of physics in terms of large Newtonian particles that act like baseballs. We can see them and their behavior. But when we study particles smaller than the microscopic we can only describe their behavior with quantum mechanics. We should be thankful that quantum mechanics governs the base of our reality. Many gross dislocations in the fertility of the universe would result in a reality that was based on the classical Newtonian mechanics of particles rather than on the quantum mechanics of waves. These dislocations would occur in all sub-universes of any metauniverse that did not rely on quantum mechanics as its fundamental dynamic. Therefore the wavey nature of reality is a stepping-stone that cannot be randomly generated by the force of a huge number of sub-universes. So the anthropic principle does not seem to explain the existence of quantum mechanics as the fundamental dynamic of reality. This remains a mystery. Anthropic reasoning would have to call for a super-metauniverse in which a minority of metauniverses operated according to the rules of quantum mechanics.

One of the most important features of quantum waves is that they automatically resist compression into a smaller space. If it is trapped in a box, it pushes out on the walls of the box. The more energy it has, the harder it pushes. If the walls of the box are moved in, the wave length decreases. In most quantum systems, a shorter wave length corresponds to higher energy. Analogously, the pitch of a trombone rises as it slides in and shortens. The higher energy of short waves pushes out on the walls harder. This effect makes solids and liquids nearly incompressible. It also prevents stars from collapsing into neutron stars and black holes prematurely. More is said about this in the next chapter.

Another feature of quantum mechanics, also involving incompressibility, is known as the Pauli Exclusion Principle. This principle states that the stable particles called fermions, that we are all made from, cannot occupy a quantum state (e.g. a region in space with a specific energy) that already has an identical particle in it. If they could, then solids and liquids would be much more compressible. The only stable states of condensed matter would be tiny super dense globs with a much greater tendency to gravitationally collapse into black holes.

This principle not only prevents implosions of matter, it also allows atoms to stick together as molecules. In atoms electrons are quantum waves. Their energy lowers as they can spread out a bit. This happens as they jump between empty states in adjacent atoms. This draws atoms closer together (as molecules) causing the electrons to lower their energy more often. If there were no Exclusion Principle, then electrons would lower their energy by condensing into the lowest state in their home atom. When other atoms approached they would form an extremely dense metallic type bond. Everything would promiscuously stick to everything else. Complex molecules and all that we call chemistry would be absent. As we have been told on TV many times: "Without chemistry, life itself would be impossible." Therefore we can consider the existence of the Exclusion Principle to be another important stepping-stone.

The Exclusion Principle was first proposed in 1921 in order to explain the colors of light that hot atoms emit. Red light from Neon signs is one example and orange light from sodium lights on the freeways at night is another. Many of the colors that scientists thought should be possible from a given excited atom had never been seen. The paradox was resolved with the conjecture that no two electrons in an atom could be in the same quantum state at the same time. Electrons have to pile up in higher and higher states as they attach to an ionized atom in order to balance the charge of the protons in the nucleus. Colors of light associated with filled states were then not allowed. Experiment and theory finally agreed. Hence the Exclusion Principle was born. No violation has ever been seen. It later became clear that this principle was responsible for the way that atoms stick together to make complex molecules.

For many years the relationship of the Exclusion Principle to other physics was not known. It was a mysterious fact of nature but the subsequent invention of Quantum Electrodynamics made it possible to derive it from four other existing principles that scientists hold dear. The derivation depends on assuming that: quantum mechanics is the dynamic of our reality; the speed of light is the same for all observers (i.e. relativity is correct); traveling backwards in time is not possible and; CPT Symmetry is never violated. Though CP symmetry is fortunately violated, CPT symmetry still has bed rock strength. CPT symmetry means that the system will look the same if the charges of particles are flipped (C) ; the system is observed in a mirror (P); and the direction of time is reversed (T). CPT symmetry would be violated if the colors of light from the antimatter version of hydrogen (or any other atom) differed from that of normal matter. Though this particular experiment is still in the near future, CPT symmetry has been well tested in other ways.

The fact that proof of the Exclusion Principle requires four pillars to stand on seems to qualify each of these pillars as stepping-stones in their own right. The compound probability of all of these things being the way that they are is miniscule. No matter how the details are tallied, the whole picture is very improbable if it is due to a sequence of chance events.

Another way that life might have failed to evolve hinges on the stability of the nuclear matter that was produced in the Big Bang. Recall that in order for a surplus of matter to have been generated in the GUT era of inflation, there had to be a way for the high energies of that time to have created an excess of quarks over antiquarks. The CP asymmetry explains in part this possibility. However, in quantum mechanics, all roads carry two way traffic. If a reaction can go one way, then it can go in reverse. During the GUT era the temperature was high enough for reverse reactions to happen as well. The excess of quarks converted back to energy and then back to a quark excess. The cooling of the universe froze in the excess. In our era of low temperatures two factors keep the energy of any proton's three quarks from reverting back to free energy. The decay of protons is inhibited by: the phenomenally high energy threshold associated with the creation of the GUT particle that can affect this transformation; and the very low probability that a proton's three quarks will have a truly simultaneous three-way collision that can make them available for this improbable transformation (to the GUT particle) to happen.

As mentioned above, all GUT theories imply that protons are slightly unstable. The earliest versions of GUTs estimated a proton average lifetime to be a trillion trillion times the age of our universe. Despite the predicted infrequency of this occurrence, experiments have been set up to catch protons in the act of dying. An early detector was in the Morton Salt mine in Cleveland Ohio. The depth of the mine helps screen out cosmic rays. A ten thousand ton tank of water, enough for ten billion trillion trillion protons, has been monitored with radiation detectors since the early 1980's. No proton decay has been detected yet. One of the earliest versions of a GUT had predicted that there would be ten decays by now. But a newer, more complex theory (Super Symmetry for example), predicts protons with greater longevity.

Just because the proton is extremely long-lived, we should not take its stability for granted. I certainly count its long life as one of the stepping-stones. If the temperature at which the GUT transformations of matter occur had been twenty thousand times lower, then the proton life span would have been reduced by a million trillion. The vast majority that were produced in the Big Bang would still be here today but things would be more volatile. For example, the Earth would be molten due to the energy from the decay of protons within. The really tough part would be that the decay of protons around us and in our bodies would produce a lethal dose of radiation. It is unlikely that the genetics of a complex being like ourselves could evolve in such an environment even if we could evolve something like the durability of the insects, which can take a lot of radiation, but they are also relatively simple.

While we applaud stable protons, the same cannot be said for neutrons. Fortunately, neutrons are stable only when bound up with protons. In their free state neutrons have a half life of a fleeting sixteen minutes, after which they decay into a proton, an electron, and an antineutrino. If a neutron's mass fell below that of a proton and an electron combined, then it could not decay. The neutron's mass is 0.1% higher than the combined mass of the daughter particles in the reaction, giving it a nice instability.

The trouble with perfectly stable neutrons is that they would have emerged from the Big Bang with the same abundance as protons. They would then have cruised around and attached to protons converting those protons into *deuterons* (hydrogen nuclei with a proton and a neutron). Hundreds of millions of years later, stars would be very short lived deuterium burners. This is because deuterons can combine easily to make helium without the need of the weak force to convert a neutron into a proton. Without this throttling of the burn rate of stars, the evolution of star-dependent life would not stand a chance.

Amongst the imaginary universes in which the neutrons are lighter than 100.05% of the proton mass, there is an even more definite point to raise. If all possible masses had equal probability, then the vast majority of these universes would have neutron masses that were less than 99.95% of the proton mass. In these universes the free protons would be the unstable variety. Therefore there would be absolutely none of the free hydrogen that provides the slow burning fuel for stars.

Before moving on to the dynamics of stars, we should stop and appreciate the strictly three-dimensional nature of our reality. Two dimensions would support the stable orbits needed for life, but complex neural circuits would not be possible. And how about four dimensions? In our three dimensional space, planets orbit according to Newton's law of gravity: force is proportional to the inverse square of the distance, i.e. as distance is divided by two, force is multiplied by four. But in a four-dimensional space (with time as a fifth dimension), force would be according to the inverse *cube* of distance: halving the distance would increase force by eight. Newton understood that only an inverse square law would give stable orbits more than three hundred years ago and highlighted it as an example of divine providence. Two centuries later, the Reverend William Paley used this observation to support his argument for the existence of the "invisible hand" of a "Master Watchmaker".

And yet, extra dimensions still might exist even as they do not impinge on our reality. An extra dimension was first proposed by T. Kaluza in 1919 in order to explain electricity and magnetism. Einstein reviewed his paper for two years and finally recommended that it be published. It attracted little attention and Kaluza died impoverished and obscure. But in the early 1970's there was a revival of Kaluza-Klein theories in the form of Superstring and Supersymmetric String theories. These more complex TOEs attempt to reconcile the four major forces: strong, weak, gravitational and electromagnetic.

The concept of extra dimensions is a challenge to visualize. Imagine that you are stuck in a deep hole. Even though you live in a three-dimensional world you have only limited access to these dimensions (directions). You can go up only as high as you can jump and you can go to the left or right only as far as the walls will allow. Similarly, forward and backward are also constrained. In string theories, the size of the hole or how high you can jump is the Planck size (e.g. 1/1,00...0 of an inch, with 34 zeros). Now if you start digging in a particular direction, the hole becomes a trench. Then you have greater access to one of the dimensions. If the length of the trench were dug to the length of the universe, then you would think that it was a completely free direction. Our three dimensions, in this respect, is like a three-dimensional membrane. Inflation blew up the three spatial dimensions that we have to much larger than the visible universe. The extra dimensions did not inflate. They remained in their compacted state at the Planck size.

Though three dimensions feel very natural and providential to us, it is not at all the right number for string theories. These recent Theories of Everything, reconcile gravity and quantum mechanics by postulating that a larger metauniverse has at least *ten* dimensions! Only three spatial dimensions and one time dimension are manifested in our reality. The rest of the dimensions are compactified. The various TOE theories differ in the number of extra dimensions and in the geometric relationships between them [1]. The particles and forces of our reality are due to the shimmering vibrations of strings in a highly contorted, super microscopic, six-dimensional subspace. In a simple string theory, the shape of this space—at each point in ours—would be a closed sphere or a donut. But neither of these shapes are complex enough to account for our improbable universe. A donut-like shape with three holes might account for the three families of quarks and leptons (the first family consists of the up and down quarks, the electron, and its neutrino). The descendents of humanity may have to ponder potential shapes for billions of years before the right one generates a TOE that is as scientifically solid as Einstein's theory of General Relativity. Nonetheless, with the possible exception of a new class of theories called Quantum Loop Gravity [2], the unification of gravity and quantum mechanics appears to require that a metauniverse keep these extra compactified dimensions. Fortunately, three dimensions are free and so we are free to pursue life.

Chapter 4
Star Light Star Bright

When we look up into a crystal clear night sky we can see our Milky Way Galaxy as a white haze that stretches across the entire sky. Only several thousand of its billions of stars are close enough and bright enough to be seen by eye. The very bright ones are short-lived giants that are good at cooking up the elements of life. Their hundred-million-year lifetimes are too short for the evolution of life. Many aspects of our universe though, allow it to be teaming with the kind of stars that could support sentient life.

In our clear night sky we can just make out the bright center of the Andromeda galaxy two million light years away. It is the nearest large neighbor to our own galaxy (our best telescopes can see galaxies that are thousands of times further away). By the way, in a billion years Andromeda will pass right through the Milky Way. The near passage of some of its stars will send billions of comets from the Oort cloud toward the Earth. Perhaps the Earthlings of the time will be able to avoid the fate of the dinosaurs if they are sufficiently beyond our level of technology (i.e. about half a century more advanced).

Anyhow, the universe today contains trillions of galaxies and almost a trillion trillion stars. The first problem we encounter when considering how stars formed is the "Lumpiness Conundrum." The universe was at one point like a smooth liquid which, as it expanded, became "lumpy" as if clusters of stars precipitated out of a clear solution. Isaac Newton proposed that stars had formed from a primordial lumpiness. But the infant universe was very smooth and stars formed too early to be explained by the observed degree of primordial lumpiness of ordinary matter. Theorists are still struggling to understand exactly how these lumps compressed with their own gravity to form stars in only two hundred million years. In the standard model of cosmology, hypothetical cold dark matter gives the process a head start during the plasma era prior to the formation of hydrogen atoms. This kind of matter could start to clump together earlier than ordinary matter because certain hypothesized particles would not interact with the plasma which was resisting gravitational compression for three hundred thousand years.

The best candidates for these electrically neutral particles are called weakly interacting massive particles. That is WIMPs. Though the acronym suggests a certain weakness the importance of WIMPs should not be underestimated. Some supersymmetry theories hypothesize massive partners of the neutrino called *neutralinos*, with just the right amount of mass and abundance to account for the observed formation of stars and galaxies. In addition, these WIMPs appear to account for most of the mass that helped produce a flat and long-lived universe. Therefore WIMPs, collectively known as dark matter because they produce no light, are probably another stepping-stone to our existence. They have not been included on the list, however, because there is still no direct evidence for them.

There is a great deal of indirect evidence of dark matter, though. For instance, the orbit of the Magellanic Clouds around our Milky Way galaxy requires that the Milky Way possess much more mass than can be seen. Likewise, the orbits of galaxies in clusters requires more mass than can be seen. The temperature of plasmatized gas trapped in clusters implies more mass than can be seen; and the velocity of stars orbiting in the galaxy's outer reaches require a heavy Milky Way. In most cases, dark matter would seem to constitute ten times more mass than the visible stars.

Some of the invisible matter is likely to consist of MACHOs which stands for Massive Condensed Halo Objects. MACHOs include mundane stuff like black holes, small dim stars, brown dwarf stars that never caught fire (like Jupiter), clouds of gas that are too uniform or sparse to form stars, the neutron star cinders of giant stars, and the cooled-off white-dwarf cinders of sun-sized stars. Evidence for MACHOs has been found in the flickering of the light from stars due to gravitationally induced optical distortions caused by passing MACHOs. But the combined mass of observed MACHOs is still insufficient to explain the effects listed above. Apparently, even though they are too shy to be seen, WIMPs are more important for our existence than MACHOs.

The first clumps of ordinary matter that started to pull together into stars consisted of enough hydrogen gas to form ten million suns. But only a small fraction of this material ended up in the first stars. This gas was drawn into the infant clumps of cold dark matter that had been coalescing for the first three hundred thousand years. In general, these clumps had some degree of rotation induced by gravitational interaction with neighboring clumps. Over a period of two hundred million years these regions of high density continued to contract and compress. As they got smaller, any rotation that they initially had caused them to separate into swirls of gas that were still part of the larger formation. This process continued even within the young galaxies to form clusters of stars and finally stars.

As a cloud of gas contracts under its own gravity the gas in it heats up from the process of compression. This happens in the cylinder of your car engine during the compression part of the cycle. The rise in temperature and pressure resists the compression. Likewise, the process of star formation slows down as pressure increases until the heat can be dissipated as thermal radiation and by boiling off some of the gas. Also, as the volume contracts the rotation speeds up, just as an ice skater spins faster in a pirouette by pulling in her arms and legs. As the rate of rotation of a star increases, centrifugal force resists the contraction in two of the three dimensions, yet along the polar axis of the system the contraction proceeds. The initially nearly spherical cloud flattens out into a disk with a central bulge. This central bulge receives more gas falling in along the polar axis unimpeded by centrifugal force.

The central bulge in the disk of gas continues to contract into a protostar while the outer disk regions break into rings like those around Saturn. Each ring becomes either a planet, if it is as small as Jupiter; or a companion star if it is ten times bigger than Jupiter and can therefore catch fire. Or, if the rings are made out of dust they can persist as rings. This is the case with the Asteroid Belt between the orbit of Jupiter and Mars. Any gas in the disk is eventually drawn into planets or into the central bulge of the star.

For a long time now, astronomers have believed that most stars have companions or planetary systems. In many cases the companion stars are visible as two separate stars that are far enough apart to distinguish them in a telescope. If they are too close together for this, all is not lost. In many cases the periodic variation in brightness of a binary star system is a give-away that one star is moving in front of the other. In fact, planets as small as Jupiter have been detected by the shadow that they cast as they move in front of their star. If the companion is too small and dim to be seen and it does not eclipse the larger star then it is still detectable by its gravitational tug on the larger star. For nearby stars, this tug is detectable as a wobble in the path of the star across the sky. Planets around far-away stars can be detected by the color shift (Doppler effect) that is caused by the wiggle of the parent star (induced by the planetary tug). At this time, evidence for over a hundred planetary systems has been found.

As the protostar in the central bulge of an infant solar system continues to contract, its center becomes hotter and hotter from the gravitational compression of the gas. As in a diesel engine, the compressive heating eventually causes ignition. This occurs at a temperature of about 500° C in a diesel engine. In a protostar, ignition occurs at about five million degrees. At first the star burns only deuterium into helium. This deuterium fuel is left over from the first three minutes of the Big Bang. It consists of a single proton stuck to a neutron. The process by which it burns is shown in Figure 4.1.

Deuterium burns readily into helium so it is consumed too quickly to support the evolution of life. If an infant star is at least one tenth as massive as the sun it will ignite the burning of hydrogen into helium. The first stars that formed were hundreds of times more massive than the sun. These giant stars burned up quickly and in their explosive death throes they seeded the universe with the complex matter that we are made from.

The ability of any star to burn hydrogen into helium is a significant stepping-stone to a fertile universe. It is a very tricky process that could easily fail to happen or it could happen too easily. Our sun is burning hydrogen at just the right rate. At four and a half billion years old, it has another four billion years to go. When it has used up all of its hydrogen, it will start to burn helium. When it does that, its years will be numbered to the millions. As it goes through its final years of burning, its atmosphere will expand out past the orbit of Earth. It will be quite a gory scene if a future civilization of human descendants, or the descendants from some other species (elephants or rats perhaps) does not get its act together and take a hike to a younger star. Perhaps an interstellar Noah's Ark will contain embryonic seed material from the entire ecosystem [1], assuming that we don't destroy it first.

Of more immediate concern than the demise of the sun, is that it is gradually getting hotter. When it first started burning, it was 25% less bright than it is today. In a billion years it will be hot enough to set off a run away greenhouse effect on Earth. If the thermal input to Earth is increased too much or if there is too much carbon dioxide in the air, evaporation of water from the ocean would add enough greenhouse effect to create this run-away situation. The extra heat kept in by the water vapor would evaporate even more water, trapping more heat. Eventually the ocean would boil away and Earth would look like Venus, with a super dense atmosphere of greenhouse gas (water here and carbon dioxide there). If we and our descendents are wise this will not happen. In the distant future it would be well within technical capability to mitigate this type of problem.

The knife-edge nature of Earth's climate is illustrated by the cold young sun paradox. Paleo-climatologists have been trying to understand why the ocean was not a block of ice when the earth was very young and the sun was cooler. The present mean temperature of Earth is only ten degrees Centigrade (above freezing). If the early atmosphere of Earth had the same thermal properties it has today, the mean temperature would have been around minus five degrees C. This would cause the ocean to freeze over completely. Once it had done this, the high reflectivity and lack of clouds would keep it that way forever. In order to explain the fact that this did not happen, it is necessary to propose that greenhouse gasses like carbon dioxide or methane were in greater abundance then. In fact there is strong evidence that the Earth did freeze over, perhaps completely, many times [2]. The build up of carbon dioxide (emitted by volcanoes) eventually led to a thaw and a super hot (e.g. more than 150 degrees F) greenhouse era.

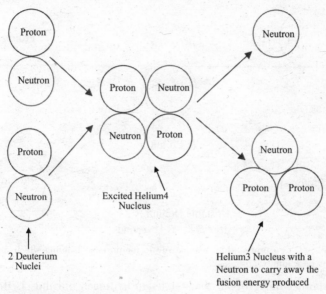

Figure 4.1 The fusion of two deuterium nuclei in a star to produce a helium3 nucleus and a neutron that carries away the energy. The helium3 later burns with deuterium into helium4 and a proton (hydrogen).

The fact that the sun can burn hydrogen for billions of years depends on hydrogen's ability to burn slowly. If the strength of the nuclear force holding the nucleus together were stronger (by only about 1/2%), then two protons would stick together to form helium2. Helium2 exists in our universe only as a resonance interaction of two protons which bind together for a very brief time, then fly apart again. This is illustrated in Figure 4.2 below. If two protons stuck together permanently, then hydrogen would burn as fast as deuterium does, and a star like the sun would use up its hydrogen fuel quickly and go into helium burning—all in a period that can be counted in mere millions of years. No life with the potential to escape its fiery fate could evolve in so short a time.

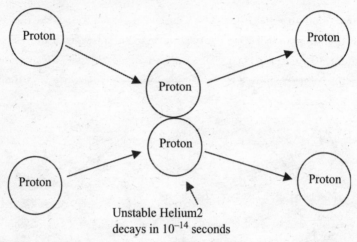

Figure 4.2 Two protons fuse temporarily to for unstable helium2 which quickly decays back into two protons.

Another lucky break that relates to hydrogen burning, is that it happens at a lower temperature than helium burning. If it happened at a higher temperature, then helium burning would dominate and again the sun would be short-lived. The slow burning of hydrogen keeps the interior of the sun inflated and at modest pressure, thus staving off the day when the violent burning of helium takes over. There are two different ways that the burning of hydrogen could have been much slower than it is. One way would have been for the helium2 resonance described above to be at a higher energy than it is. This would have occurred if the strength of the nuclear force (the strong force) were a little weaker than it is or if the charge on the electron (the strength of the electromagnetic force) had been significantly stronger. However, both of these changes would probably also alter the burning of helium. The result might be stars that don't burn at all or burn only if they are huge. These would be short-lived. They would either blow up quickly or cool down quickly.

Yet another way for the burning rate of hydrogen to be changed or killed off all together is for the strength of the weak force to be different than it is. In order for hydrogen to burn into helium it must first produce deuterium. Then, as mentioned above, the deuterium readily burns into helium. The production of a single deuterium nucleus from two protons (2 hydrogen nuclei) is illustrated in Figure 4.3 below. This reaction requires that one of the protons be turned into a neutron, an antielectron and a neutrino by the weak force. During the brief instant that two protons are whipping around each other as unstable helium2 one of them can turn into a neutron that can bind to the other proton. The binding energy that is released is enough to more than pay back the energy debt that is incurred by changing a proton into a neutron, an antielectron, and a neutrino. But most of the time the two protons split apart again without one of them converting to a neutron. The stronger the weak interaction is, the greater the chance for the conversion. Thus the rate of burning, in sun like stars, is throttled by the weak force.

If the weak force were significantly weaker than it is, then many more proton-proton encounters would be needed to produce deuterium. This would require the higher temperatures at which helium starts burning. This in turn would result in a star that started its short life burning helium or that did not start burning at all. We would have been denied our moment in the sun. Or the moment would have been too brief for bacteria to make much (e.g. us) out of it.

A way that hydrogen burning could have been too fast is if a neutron were lighter than a proton plus an antielectron. This was also discussed in the previous chapter. In this case the Big Bang would have produced a super abundance of heavier elements. In our universe neutrons are unstable and decay into protons. In this alternate universe the protons would be the unstable particle. Any hydrogen nuclei (protons) would decay into neutrons that would then wander around looking for a nuclei to attach to (to make deuterium). But there would have been no hydrogen in the aftermath of the Big Bang. So later, any stellar burning would be the burning of deuterium or helium. As pointed out above these would have been fast reactions that would consume stars in an instant on an evolutionary time scale. Heavyweight neutrons have allowed our lucky stars.

Let's return now to the universe that we do have. The life of most stars proceeds in a state of relative stability as the supply of hydrogen in their cores is gradually consumed. The larger the star the more rapidly this happens. The smallest stars, called red dwarves, are about one tenth the mass of the sun. They burn so slowly that they can last for hundreds of billions of years. One of the nearest stars to us, Barnard's star, is like this. Though it is one of the three closest stars to us it is not visible to the eye.

At the other end of the spectrum of brightness are stars like Sirius, the dog star. Even though this star is much further than Barnard's star it is the brightest in our sky. It is 20 times more luminous than the sun and about two and a half times more massive. Because of the intensity of its nuclear fire, it will use up its core hydrogen in a billion years. It is unlikely that its solar system will produce complex organisms in time for them to save themselves from their fiery fate.

Stars more massive than Sirius live for even shorter times. They proceed through a sequence of changes that culminate in a supernova explosion. Of course, because of their short lives they will not be able to host life directly. However, stars like this are very important to the ability of this universe to produce life because they are the furnaces in which the elements of life are forged. When they blow up, elements like carbon, oxygen and nitrogen are disbursed into space. Later, this debris can gravitate together to form second generation stars like our sun and it is in their solar systems where rocky and wet planets like Earth can be the cradle of life.

When a star uses up its core hydrogen it starts to burn helium4 (2 protons+2 neutrons = 4 nucleons) into carbon12. Close examination of this process reveals yet another stepping-stone. In order to make carbon, three helium nuclei must come together almost simultaneously. This is because the nucleus that results from the fusion of two helium nuclei is the very short-lived resonance called beryllium8 (4 protons + 4 neutrons = 8 nucleons) . The naturally occurring kind of beryllium (beryllium9 = 4 protons+5 neutrons), is quite stable but beryllium8 is not. But without beryllium8 the production of carbon

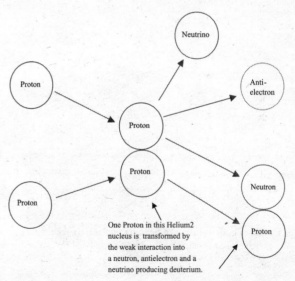

Figure 4.3 Two protons fuse temporarily to form unstable helium2 but one of them is transformed into a neutron, by the weak interaction, thus forming stable deuterium (hydrogen2). This is the first step in nuclear synthesis in the sun.

would occur through the very improbable three body collision of three helium nuclei. This is illustrated below in Figure 4.4.

The burning of helium would not happen to any significant degree if it were not for the existence of the beryllium8 resonance at an energy that is accessible to the helium in the core of a star. This resonance allows the two helium nuclei to stick together for 0.00000000000000001 of a second (i.e. 10^{-17} seconds), just enough time for a third helium nucleus to join the party and make the fun loving element called carbon12 (see Fig. 4.5). This element has the potential to congregate into huge parties of thousands of atoms called proteins. The dance of thirty thousand different kinds of proteins forms much of the basis of our cellular life.

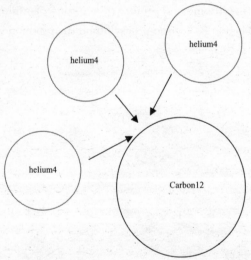

Figure 4.4 The three body collision of helium4 nuclei to make a Carbon12 nucleus. This almost never happens because of the improbability of two simultaneous bull's eye hits.

Without this beryllium resonance boost the production of carbon from helium would be negligible in the life of a star. Importantly, this reaction in stellar furnaces is the way that almost all carbon comes into existence. After all of the hydrogen had been burned into helium, stars would cool off and contract if the beryllium8 resonance were absent. The rise in pressure would eventually squeeze the star into a white dwarf, neutron star, or black hole (depending on its size) without ever seeding space with the elements of life.

A second way that the universe could have been deprived of carbon is if most of it were more easily converted into oxygen (also a fun element, but it is no fun without carbon). The conditions that allow three helium nuclei to burn into one carbon are also conducive for adding a fourth helium that rapidly converts the carbon into oxygen [3]. This reaction is so fast that there would be very little carbon around if something extra didn't speed up the production of carbon to keep it ahead of its conversion to oxygen. This something extra was predicted by the astronomer Fred Hoyle based on the hunch that since the universe has an abundance of carbon, it must have a mechanism for turbo charging the production of carbon relative to its conversion into oxygen. He hypothesized the existence of a resonance of carbon at just the right energy to form from three helium nuclei. Such a state would split apart quickly back into three helium nuclei if it did not have a way of dissipating its excess energy. If it has just the right electronic properties, it can do this by emitting an energetic burst of light called a gamma ray (e.g. the very high energy X-ray in Fig. 4.5). This would allow it to settle into stable carbon12.

A second prediction by Hoyle was that there would *not* be a corresponding resonance in oxygen16. A certain resonance does exist, but not at the same energy. The energy of a carbon nucleus plus that of a helium nucleus is higher than the energy of this resonant state of oxygen by only 1% of a typical nuclear energy level spacing. Without this energy difference, carbon12 could only result from exotic reactions such as those that make relatively rare lithium, beryllium, and boron during the collisions of supernova debris with interstellar gas. The abundance of carbon would be reduced to 0.01% of what we have now.

Both of these detailed predictions about oxygen and carbon resonances were made by Hoyle on the basis of anthropic reasoning. The emergence of carbon-based beings requires these two conditions. Since we exist, these resonances must exist at the energies required to produce carbon in abundance. Subsequent to these predictions in 1954, Hoyle managed to cajole his nuclear physics colleague Willy Fowler into investigating the issue experimentally. He was soon proven to be right. This is perhaps the greatest predictive success of anthropic reasoning to date [4].

So what's so great about carbon anyway? Aren't there many other conceivable paths to some form of life? The answers to these

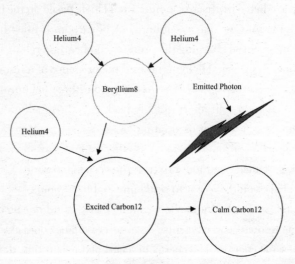

Figure 4.5 The actual way that a carbon12 nucleus is produced with the short lived beryllium8 resonance as a stepping-stone. The excited state or resonance of carbon12 is at just the right energy to catch the three helium nuclei.

questions will be attempted later on. For now let it suffice to point out that the chemistry of carbon is the most complex of any element. It can form chemical bonds with up to four other atoms. A very attractive feature of carbon is its ability to form single, double, and triple bonds. Single and double bonds are used extensively by the chemistry of life as we know it.

There may be life forms based on silicon elsewhere in the universe. But while silicon can form bonds with up to four atoms as well it can not form double bonds because it is too big. Also, silicon doesn't naturally form a stable gas. Silicon dioxide is sand. Though these draw-backs may not be fatal to silicon-based life they certainly reduce its potential.

Now let's get back into the interior of a giant star because we are not through tracing the origin of the heavier elements like silicon and iron that are necessary for planets like Earth. Once a star starts burning helium it runs through its supply in the core quite rapidly by cosmic standards. For a star that is 25 times more massive than the sun the burning of hydrogen would go on for about 10 million years. The period of helium burning would last for only 1 million years. When the core helium is used up, it starts cooking up heavier elements than carbon, like silicon and magnesium. The helium burning reactions may still go on in an outer layer of the core where the helium is not used up, however. This can occur if this layer has not already been boiled off into space by the prodigious energy production that occurs in this stage of a stars life. Deeper in the core, carbon is burned into neon for 600 years. When the carbon is used up in the core, again the reaction continues in an outer layer. Therefore, the star begins to look like an onion with neon in the core, encased by layers of carbon, helium, and finally an unburned layer of hydrogen mixed with primordial helium. Burning reactions

occur at the boundaries between the layers and as the star contracts
and the core heats up to 1.5 billion degrees Celsius (the sun's core is
11 million degrees), the neon disintegrates into oxygen. In only half a
year all of the oxygen burns into silicon. In only one day, the silicon
in the core burns into iron. This sequence is illustrated in Figure 4.6.
(Each layer in this figure is dominated by the element shown but
would also include other elements of similar mass.) But the iron cannot burn into any higher nuclei because the iron nucleus is the lowest possible energy combination of neutrons and protons. Making
heavier nuclei from it would require the input of energy instead of
the release of it. Heavier elements than iron are rare because they
are made only in supernova explosions that can supply the necessary
energy in a flood of neutrons. In a giant star the ball of iron in the
center just grows bigger without generating any other elements.

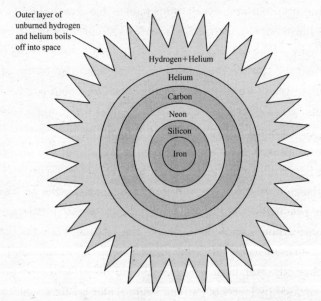

Figure 4.6 The layers of elements in a large star near the end of its' life.

But between the outer layers burning proceeds and the rate of energy production goes way up. The outermost layers of the star become so hot that they begin to boil away into space. For large stars this process strips away the hydrogen and helium layers and even a little of the carbon layer. Oxygen and nitrogen can also be found in this layer. But the amount of these elements that gets into space in this way is probably too little for a fertile universe if this were the only way to get them out of stars. The heavier elements like silicon, which makes rocks and phosphorous, which is needed for DNA, remain for the most part deep inside of these giant stars.

To disperse these higher elements stars must blow up in supernova explosions. These explosions happen when the iron core gets too large to resist the compression of gravity. Because iron cannot release more energy by burning into higher elements, the core cannot make itself hot enough to accommodate the pressure of its own weight. A point of instability is crossed called the Chandra limit when the core exceeds about one-and-a-half solar masses. Then the bottom literally falls out from under the outer layers of the star. This happens so fast that these outer layers remain unperturbed like a cartoon character that has run off a cliff but doesn't fall until he looks down and panics.

Once the collapse of the iron core begins, it only takes a few seconds for it to plummet into a ball of nuclear matter that is only ten miles in diameter. Protons in the iron nuclei change into neutrons because the Exclusion Principle makes the presence of electrons (that balance the electric charge on the protons) energetically costly. Electrons are squeezed into the protons converting them into neutrons and creating a tremendous flood of neutrinos. The growing neutron ball at the center of the implosion is compressed to incredible pressures by the impact of the in-falling material and, in a violent recoil, it blasts away its outer layers. The implosion is now an explosion. Meanwhile, back in the remains of the core, the neutrinos are depositing a huge pulse of energy into the collapsing iron. The energy from the neutrinos superheats the iron and creates a buoyant bubble that is driven upwards like a thunderhead on a hot day. This happens just as the shock wave from the recoil of the ball of neutrons at the center hits this super-heated core layer. The combined effect of these two sources of energy pushes the shock wave towards the surface of the star despite the huge amount of iron that is falling in. This process is illustrated in Figure 4.7.

All of this happens in the space of a few seconds. At the surface of the star there is still no hint that anything has happened. A day later the shock wave hits the surface which is much further away at the distance of Jupiter's orbit from the sun because it has been swelling from the high rate of energy output of the previous few hundred years. When the surface is breached by the explosion, the supernova then begins to become apparent to all within hundreds of thousands of light years. The outer regions of the iron core are blown out into space. The tremendous flood of neutrons builds heavier elements up well past uranium. The afterglow of the explosion has the characteristic decay time (thirty six hours) of the extremely heavy and unstable element called californium . All that remains is a ball of the neutrons (that did not recoil from the implosion) called a neutron star.

A supernova explosion occurred several thousand light years from Earth in 1054 AD. It was recorded by Chinese astronomers of the day. The decay of its californium lit the night sky for several days with an intensity that was brighter than the full Moon. The spinning remnant neutron star was identified as a pulsar by its characteristic radio pulse emissions. The beautiful Crab Nebula is the dust left from the inner layers of the onion. Our earth and most of the matter in our bodies are the star dust from similar explosions that occurred between five and ten billion years ago.

In 1987 a supernova was detected in the Large Magallanic cloud of stars in orbit around the Milky Way. The most interesting thing about it was that twenty-two neutrinos from the flood that was produced were detected in three separate locations. Neutrino astronomy was born and this first observation confirmed the general picture of stellar collapse outlined above.

The fact that neutrinos can escape the neutron core and that they interact right where the shock wave needs them, transforming the implosion into an explosion, is another stepping-stone to our life. Supernovae would not explode without this effect. If the strength of the weak force (that dictates neutrino behavior) were significantly different, then supernova explosions would be duds. All of the matter in these giant stars would end up forming black holes. The kind of rocky planet that was the cradle for our lives would not form.

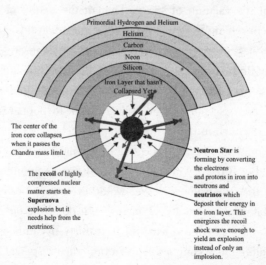

Figure 4.7 Mechanics of a Supernova explosion in the core of an old giant star.

In addition to providing the raw material for our bodies and our kind of planet, supernovae also seed it with a supply of long lived radio active elements (uranium, thorium, and potassium40). These elements keep the rocky planets that ultimately result geologically active for billions of years. In the next chapter, we will see that this geological activity is deeply intertwined with a stable atmosphere, ocean, land mass, and the evolution of complex life.

The supernova explosion described above is known as Type II. Type I supernova also seed heavy elements (but almost no carbon) into space by an entirely different mechanism that also requires nuclear phenomena to play in a tightly choreographed dance. This type of supernova produces a very consistent amount of light. By observing of a large number of these explosions, scientists have come to the startling conclusion that at this time the speed at which distant galaxies are rushing away is *increasing* rather than decreasing. As in the inflationary era, gravity has become repulsive due to "dark energy" in the vacuum of space. Though the nature of this dark energy is entirely mysterious, it is needed to explain this observation and the flatness of space seen by WMAP (The Wilkenson Microwave Anisotropy Probe is a more advanced COBE-type satellite.) Though dark energy may not have been needed to get us to this point in time it might give our descendents a great deal of time to ponder it. Because of it, a Big Crunch in the future is very unlikely. A recent report on new measurements claims to rule out an alternate doomsday scenario called the Big Rip [5] which would result from an runaway acceleration of the expansion rate that would eventually destroy atoms. Though the data is actually inconclusive, it does rule out this inflation-like catastrophe for at least the next thirty-five billion years.

The production of complex elements and their dispersal into space required billions of years. Like the layers of those giant stars from which the material of our bodies originated, the nature of matter as it has changed over time has innumerable layers. Levels upon levels churn upon themselves to beget further levels of reality. This is what it took to set the material stage for the drama of our evolution.

Chapter 5
Mother Earth

On his way back from the Moon, Edgar Mitchell looked at the approaching Earth and "experienced an ecstasy of unity" with the process of life. The deep experience forever changed the way he saw things [1]. He and his fellow astronaut saw in the black void of space this blue, green, and brown planet stand out as unique and beautiful. Water, air, and land coexist, providing the rich environment from which complex life can evolve. The geological forces that continuously reshape the surface have forced new forms of life to evolve by changing the climate and habitat. Life in turn has shaped the land, sea, and sky in countless ways. This coordinated interaction appears so organic that some even believe that it behaves as if it were a single organism called Gaia [2].

In order for this complex ecosystem to have evolved, there are numerous requirements that go beyond those that have already been discussed. These new stepping-stones to life provided the rich environment of land, sea, and air, and a state of relative climatic stability for billions of years. Though stability is a prerequisite, a certain amount of instability is required as well. Without the ten major extinction events of the last billion years, life would have evolved too slowly for any creature to have any hope of coping with the eventual demise of the sun. By tracing the history of the Earth, we can see that the evolution of life is driven by the tension between static existence and the ever-present dynamics of oblivion.

This dualism is expressed in many cultures. Examples are: good and evil; God and the devil; light and the shadows; yang and yin etc.

As pointed out earlier, the history of our planet began with the formation of an excess of protons in the inflationary era of the Big Bang. However, these protons had to be built into the complex elements of life in the cores of giant explosive stars. The raw material for Earth was thus forged about five to ten billion years ago. The supernova explosions that ended their short lives blew 90% of their material into deep space while the inner 10% collapsed into neutron stars or black holes. Also many elements heavier than iron including hundreds of radioactive elements were created when the flood of neutrons interacted with the ejected material. Three of these radioactive elements live long enough and are sufficiently abundant to be the principle source of heat for the volcanism and continental drift that has shaped this planet into the cradle of life that it is. They are potassium40, uranium and thorium. Potassium40 has been the most important source of geothermal energy up until the present. In about a billion years all three will be equal players.

The rich mix of complex elements that was blown out into space remained adrift in the Milky Way as a cloud of gas and microscopic particles of interstellar ice and dust for billions of years. Almost five billion years ago a final supernova explosion in our vicinity compressed the cloud enough to initiate its contraction. Gravity compressed a region of it into our solar nebula (as described in Chapter 4). The initially spherical nebula flattened into a rotating disk of dust and gas, and as the central bulge became the sun, outer regions of the disk self-gravitated to form the nine planets. A process initiated by chemical adhesion of ice and dust started building small dirty snow balls called comets and rocky asteroids. The heat from the radioactive decay of aluminum26 caused the larger asteroids to melt and fuse into solid rocks.

Asteroids orbiting the sun occasionally crash into each other. If they are small, they simply fragment and fly apart, but if one is larger than about a mile in diameter, then it will have a strong enough gravitational field to capture most of the fragments. As it grows, its gravity becomes stronger and it gathers material with greater efficiency. Thus, the asteroids and dust that had been scattered in the solar disk gradually coalesced into the nine planets that we know of.

Had Jupiter not been so massive, a tenth planet might have formed between it and Mars. The strongly disruptive effect of Jupiter's gravitational field kept the asteroids in its vicinity (that were too far to be captured) from condensing into a planet. These form the asteroid belt between Jupiter and Mars today. The strong gravitational field of Jupiter also cleared the vast majority of asteroids out of the inner solar system at an early stage. This provided Earth with a relatively tranquil environment. The average time between killer asteroid strikes is about a hundred million years, an interval long enough for stable ecosystems to evolve.

The sun burned fiercely at first. Its atmosphere expanded out to the orbit of Earth and seared the inner planets. Most of the light, gaseous elements were driven from the surface. They were then either gravitationally drawn into the sun or driven further out by the intense solar wind of plasmatized gas boiling off the surface of the young star. Jupiter and the other outer planets were far enough, and massive enough, to retain their gases, but retained so much gas that the temperatures and pressures at the bottom of their atmospheres precludes the existence of life. Similarly, had the Earth's initial atmosphere been retained, then it would have become a sterile super-greenhouse pressure cooker like Venus, the loss of Earth's original atmosphere was a stepping-stone its habitability.

Millions of years later, another planet the size of Mars collided with the Earth. A huge quantity of silicate rock was vaporized and formed a ring system of dust, like Saturn's, in a low orbit around the Earth. The disk of dust gradually coalesced to form the Moon. The compositional similarity between the Moon rocks brought back by the Apollo missions and rocks of the Earth's crust supports this hypothesis. Initially, the Moon condensed from the debris in a relatively low orbit and generated tremendous tides in the Earth's nascent ocean. The interaction between ocean tides, the spin of the Earth, and the Moon slowly transferred energy to the Moon, pushing it into an ever higher orbit. This transfer of energy also meant slowing the spin of the Earth and lengthening the day.

For life, the Moon is more than a nice trinket in the sky. The cycle of the tides it causes was a key environmental stimulus to our journey from the sea to the land. Take the lungfish for instance. It lives in shallow tidal pools. Its need for oxygen when the water receded during low tides contributed to the evolution of its primitive lungs. The richness of life around the continental margins, in shallow waters, regularly flushed by tides, made the land an inevitable habitat. The concentration of nutrients caused by cycles of tidal flooding and evaporation may also have been important to the initial evolution of primitive life. In addition, the Moon stabilizes the rotational axis of the Earth. Without it, catastrophic climate fluctuations would occur at intervals of tens of millions of years due to excessive tilt of the polar axis (i.e. the arctic circle would occasionally reach down to the equator). The impact that formed the Moon is another example of how catastrophe has bred opportunity for life.

After this major collision, the Earth continued to be pelted by small meteors and comets. Meteors brought in iron, silicate rock, and carbon. The comets brought in water and organic material that had formed in interstellar space. All the while, volcanoes drove gasses and water vapor from the rocks into the air. So volcanoes not only built the land but they also contributed to the sea and the air by providing most of their raw materials. In addition, it is quite possible that the earliest forms of life evolved in volcanic hot water vents deep in the ocean.

Initially, most of the heat of the Earth's volcanism came from energy delivered by crashing comets and meteors, the heat of impact liquefied the Earth. This allowed iron from the meteors to settle into the Earth's center where it formed a molten core. Lighter silicate rocks floated to the surface to form the "plastic" mantle.

The continental plates float on top of the denser mantle. The convective churning of the mantle continuously brings up heat from the depths and shoves the continents around. Continental drift is measured in inches per year. In a half-billion-year long cycle, the continents are alternately driven together to form supercontinents and then pulled apart. The fragments gather together at places where the mantle is cool and is descending toward the core. But the thick continents act like a blanket. Over a period of hundreds of millions of years, these insulated cool places become hot spots and the flow reverses. The rising and spreading plume of hot mantle then tears the supercontinents apart. Their fragments then disperse toward new regions where the mantle is descending. Where the oceanic plates are driven under the continental plates, at the continental margins, the friction generates local pockets of molten rock. These pockets of magma melt their way to the surface and form most of our volcanoes.

Our Improbable Universe

The collisions of the drifting plates and volcanoes are continuously rebuilding the land masses and renewing the atmosphere. Without these processes, the land would be eroded down to sea level by rain and waves. There has been enough erosion since the Earth formed to have done this many times over. The mile deep Grand Canyon took only 0.2% the Earth's age to erode. So for us land-loving types, the existence of geological activity has been one key to the survival of our habitat. The slow changes in our habitat have helped drive the evolutionary process that has birthed us.

However, an absence of land masses would not preclude the development of intelligent life. Whales and porpoise have developed large brains to service their sonar navigation and their social life. The octopus also has a relatively large brain in order to process the flood of tactile information from its eight arms and to coordinate their complex motion. The sea otter even uses a pair of rocks as tools to crack open its shelled meals. Despite these mental resources, however, it is unlikely that these animals will ever evolve a technological civilization like our own that has the potential to outlive the sun. With the exception of the sea otter, they have changed little in tens of millions of years. The ocean is just too stable and uniform an environment to whip them up the evolutionary ladder. The liquid environment makes the production of metals impossible. Though technology is not necessary for individual and small group psyche, it is necessary for the kind of super-collective psyche that is at the foundations of our politics, science, and arts.

Our need for an active geology is illustrated by Mars. It once had oceans and a reasonably dense atmosphere. Life may have evolved there billions of years ago. But if it did, it has failed to thrive for lack of a stable ocean. Long ago there were large bodies of water there.

Rain fell, rivers flowed and waves beat the shores of shallow seas. Today the water is mostly gone. It is almost all gone now because there is not enough heat generated in the interior to drive an active geology. The inactivity is evidenced by the lack of a magnetic field and the lack of sufficient volcanic activity.

When most of the volcanoes on Mars went extinct the fundamental source of water and carbon dioxide was lost. The ocean then contained the seeds of its own destruction. Over a period of millions of years it slowly converted the carbon dioxide atmosphere into carbonate rocks. This happens in our ocean as well. However, on Earth, the carbon dioxide gets recycled into the atmosphere when the carbonate rocks are subducted, with the oceanic plates, under the continental plates. The friction between the sliding plates, melts the rocks at the boundary. The pool of magma then melts its way to the surface where it makes volcanoes

We need carbon dioxide in our atmosphere for, among other things, photosynthesis, which produces almost all of the sugar at the base of the world's food chain. In addition carbon dioxide plays an important role in keeping our ocean from drifting into space, as it acts like a blanket over the entire Earth. In the stratosphere, it captures some of the thermal radiation that the Earth's surface would otherwise radiate into space. As air rises from the surface it expands and cools. By the time it reaches the stratosphere it is cooler that the air that is already there (because its carbon dioxide has been absorbing heat). The cooler air from below can go no further up at this point. It has hit a barrier called a temperature inversion.

Temperature inversions happen in the lower atmosphere, too. When they happen over major cities in the summer, smoggy pollution gets trapped. In the case of the global temperature inversion, it is water vapor that gets locked in. We are fortunate to have this carbon dioxide driven temperature inversion. If the water vapor could rise higher it would reach the edge of space, and ultraviolet light from the sun would break the water molecules into hydrogen and oxygen. The very light hydrogen would then float off into space. Over a period of billions of years the oceans would vanish and complex life on Earth would be doomed.

Fortunately, the Earth has had enough internal heat to keep volcanoes going for billions of years. Initially this heat came from the impact of meteors, but this source is short-lived on geological time scales. In the nineteenth century, Lord Kelvin, a pioneer in the field of thermal dynamics, asserted that Darwin's theory of evolution could not be correct because, according to Kelvin's calculations, the Earth could not be more than forty million years old, where as Darwin had stated that evolution had required hundreds of millions of years. In this time the molten surface of the Earth would have cooled down to the point of solidification to such a great depth that volcanism would have ceased. Earth's active volcanoes were taken as proof that it could not be old enough for life to have evolved.

In the eyes of the scientific community and in his own, Darwin's theory was in big trouble. He even considered retracting it. However his theory was rescued by the work of Marie and Pierre Curie. Marie Curie chose to investigate the phenomenon of natural radioactivity, which was first observed in 1896 (by another French scientist, Antoine Becquerel) and she devoted her professional life to the study of this phenomenon. (Her death from cancer might indicate that she gave her life for this study as well.)

Marie Curie set out to chemically isolate the source of the X-rays in a uranium ore called pitchblende and she was later joined in this effort by Pierre. Eventually they refined a thirtieth of an ounce of a previously unknown, highly radioactive element, later named radium, from eight tons of pitchblende. It was determined that radium was being created by the natural decay of uranium atoms.

Measurements made of the heat generated in the radium, and its rate of decay, brought the realization that a huge amount of energy was being released from the Earth's pitchblende. The nuclear energy in an ounce of uranium was roughly what could be obtained from burning thirty tons of coal. Here was a source of heat that could keep the interior of the Earth molten for billions of years.

Darin's theory was saved, but the nuclear age was born. Two years later, in 1905, Einstein published his famous equation: $E = mc^2$. This equation is commonly credited with being the first indication of the magnitude of this terrible nuclear potential. However, the previous experimental work described above had already released the Genie in the Bottle of the nucleus.

Lord Kelvin's estimate of the Earth's age was also wrong because he assumed that volcanoes required a shallow molten core. He did not know that solid rock was plastic enough to convect up from great depths. Some of Earth's geothermal energy remains from its violent formation, but without the contribution from natural radio-activity, Earth would be very much less volcanically active. After most of the initial store of heat from meteor impacts had radiated into space, the decay of radioactive nuclei became dominant and has helped keep it geologically active ever since. Though this nuclear energy is a curse when earthquakes knock down your house, or lava flows drive your species into extinction, or when it provides human-ity with the means to self-destruct, it has been necessary for our complex environment.

Only three kinds of radioactive nuclei contribute significantly to the heat of the earth today. All of the others decay too rapidly, too slowly, or they are too rare. The ones that are just right are potassium40, uranium, and thorium. Potassium40 contributed the vast majority of the earth's geothermal energy since its formation. Because it decays more rapidly than uranium and thorium, its contribution is beginning to wane. Presently it is estimated to contribute about the same amount of heat as uranium and thorium combined.

The right lifetime and abundance of at least one radioactive species is one of the stepping-stones to life here on Earth, but it is not a very improbable one in that there are so many radioactive candidates for the geothermal role. Also, a sufficiently large planet or a relatively speedy evolutionary process could have produced curious creatures before volcanism ceased even if there were no natural radioactivity. In addition, gravitational interactions, such as those that keep the Jovian moon Io molten, can provide long-term heating. So there are other ways of getting sentient life in a Nuclear-Free Universe. Nevertheless, Earth has relied on potassium40 to keep its interior hot.

The active geology of Earth plays a role that goes beyond its involvement in the maintenance of the atmosphere and the oceans. It is a crucial force that drives the evolution of complex life. The constant slow changes of the land forms have caused changes in climate and sea level that have forced many biological adaptations. Mass extinctions caused by geological changes have opened up biological niches that have occasioned evolutionary adaptation. Roughly fifteen million years after each such event, animals and plants adapt fully to the new environments, and stable ecosystems develop. A period of great stability follows and evolution slows, until the next major environmental shift. Without the turmoil of an active geology (and meteor impacts, see Chapter 7), life on Earth would have stagnated and we might be as stupid as the timeless alligators. When the sun was finally ready to blow up, no one would know that it was going to happen or know what to do about it.

An example of geologically influenced evolution may be the Permian extinction. This was the greatest extinction to occur since complex life evolved. It wiped out 96% of all species. The events that may have led up to this extinction started with the joining of all of the continental land masses into one huge one that we call Pangea. This ultimately led to a poisonous up-welling of stagnant water from the deep ocean that killed off many of the species that lived in shallow waters a quarter billion years ago. Very soon after, a plume of molten rock, from 2000 miles down in the earth, broke through to the surface. It produced a massive flow of lava in what eventually became Siberia, which covered an area more than a thousand miles in diameter in less than a million years. The released sulfurous gasses and ash probably contributed to the extinction of many of the reptiles and amphibians of the day. Importantly, the spurt of evolution that occurred in the wake of these events led to the ancestors of mammals and the dinosaurs.

From this example, it can be seen that the dynamic thing we call life needs an environment that balances stability and change. Periods of stability are used to absorb the changes. They allow the evolution of new life forms that are adapted to the new environments. Periods of radical change clear the decks and create new biological niches. A huge amount of time and many cycles of change were needed before evolution reached the point of producing knowing witnesses. Amazingly, the universe now contemplates itself through the eyes and minds of creatures like ourselves.

As we saw in earlier chapters, many stepping-stones had to be correctly positioned before even a solar system could come to be from star dust. This star dust had to have enough long-lived radio-activity to keep our planet churning and bubbling for the four billion years that we took to evolve. The right amount of internal heat and the right distance from the sun kept us from being in either a frozen waste-land like Mars or a simmering cauldron-like Venus.

All scientific evidence supports the belief that the laws of physics are the same throughout the part of the universe that we can see. Therefore it is to be expected that the conditions for life could occur elsewhere because there are so many stars and solar systems that are like our own (trillions of trillions). Many centers of intelligent life probably exist in our universe.

In addition to the universe-wide phenomena that were necessary to create the stage upon which the drama of life on Earth could unfold, there is a plethora of hurdles that were specific to the evolution of life on this planet. If all of the creatures that are our ancestors took twice as long to evolve or failed to pass a single major barrier, then it would be the absolute fate of all life on Earth to end when the atmosphere of the sun expands out to Earth's orbit. So in the same way that the story of life in the universe depends on many physical stepping-stones, the story of life on Earth depends on many biological ones. Now we will explore some of these biological stepping-stones to intelligent life on this planet.

Chapter 6
Muck Makes Microbes

Just how the complex chemical environment of our infant planet produced life is still obscure. Many creative hypotheses have been proposed but none have acquired the degree of experimental support necessary to develop a scientific consensus. As with the details of the Big Bang, the lack of ancient evidence leaves a lot of room for speculation. However, as with the Big Bang, there are fragments of information which allow enough perception to appreciate the magnitude and complexity of the occurrence. There is now enough knowledge to see some of the biological stepping-stones to self-reflective life.

The oldest evidence of life is found in rocks that are 3.8 billion years old [1]. This is only a few hundred million years after the oldest rocks formed after the Earth's surface solidified. This amount of time is less than 2% of the age of the universe and the expected lifetime of the sun. Therefore the initiation of life here seems to have been a very probable event even though the substructure of our universe is not. If it hadn't happened in the first several hundred million years there were still billions of years left in which it could happen. In that solar systems are very common, at least primitive life must be common throughout our remarkable universe. Therefore this discussion of the history of life on Earth is also a universal history in a broad sense.

The assertion that biological life is a common phenomenon is supported by more than its rapid development here. The molecules upon which our form of life is based commonly occur in space and meteors. The Murchison meteorite was found to contain the building blocks of proteins, known as amino acids in 1969. These amino acids were not created by biological processes but simply arose from the complex chemical mix in the meteor in its environment of deep space. Formaldehyde and numerous other organic compounds have been detected in the Giant Molecular Clouds in deep space. These clouds are the nurseries for infant stars. Organic material that can form cell-like membranes has been extracted from meteorites [2]. The comets that exist in great profusion in the solar system and its outer most regions (the Oort Cloud) are dirty snow balls that consist of a mixture of ice and many different hydrocarbon molecules.

The ability for the molecules of life to spontaneous form in simple chemical systems was first demonstrated in 1953 by Stanley Miller. Following up on suggestions by his mentor, Harold Urey, he subjected a mixture of gasses that are common in our solar system (methane, ammonia, hydrogen, and water vapor) to an electric discharge (ultra-violet light from the sun works too). After one week, the resulting yellow-brown gooey mess of resultant chemicals was analyzed, and it was found to contain large amounts of amino acids. Twenty different amino acids are the building blocks for all of the proteins that are a major pillar of our biochemistry today. The chemistry of the human body is based on about thirty thousand different proteins. Many of these proteins contain tens of thousands of amino acid building blocks. The simple intestinal bacteria known as E-Coli is based on about five hundred different proteins. A minimum of a couple of hundred appears to be necessary for self sustained cells.

Though the Miller experiment demonstrates how easy it is to create some of the molecules of life, from simple and common mixtures, it has two major problems as a model of the genesis of life on Earth.

One modern theory of the Earth's early atmosphere does not allow for very much ammonia or methane. These compounds are common in the outer solar system (Jupiter and beyond) but were destroyed in the inner solar system when the sun was first burning fiercely. After the sun settled down, our atmosphere formed from gases driven out of the rocks (nitrogen, carbon-dioxide, and water vapor) by volcanism.

Some propose, it is still possible that life first formed in the outer solar system, or even in another solar system, and was transported here on meteors or comets [3]. Pieces of Mars can be found as meteorite fragments on the Antarctic ice sheet [4]. Undoubtedly material from the moons of Jupiter (where methane and ammonia can be found) will eventually be identified as well. The lack of ammonia and methane in the Earth's early atmosphere does not exclude something like Miller's experiment from being a stepping-stone to life on Earth. Its origins may simply be elsewhere in the solar system or even in the galaxy.

Another problem with placing this type of chemical system at the genesis point is that protein based chemistry may in fact be a relatively late development in the history of life. Proteins do not have the very fundamental ability (fundamental to all life) to self-replicate. Self-replication is probably all that the first forms of life did. Crystals self-replicate their structure by positioning new atoms on the exposed surface in exactly the same structure as the underlying ones. The surface serves like a template that organizes the next layer of atoms into the same structure. However we do not call crystals life because they usually do not have the ability to become more complex than they are. Mistakes such as the incorporation of an "incorrect" atom tend to be covered up rather than propagating into later layers. However, the growth process of particles of certain kinds of clays has been identified as having the ability to propagate defects, induced by organic molecules, from old layers into new layers [5]. The hypothesis that life started this way is considered to be a long shot for the source of all life by most knowledgeable reckoning.

Two types of molecules capable of self-replication and mutation, are the basis for all life today. These are molecules of RNA and its cousin DNA. Both these molecules consist of long strings of sugar-like units linked together by phosphate groups. The sugar units in RNA and DNA are slightly different. Four different units called "bases" are attached to this string of sugars. This structure is depicted in Figure 6.1. The sequence of bases can be thought of as the writing on a ticker tape in which only four letters are used. Every nucleus in human cells has a one yard length of DNA curled

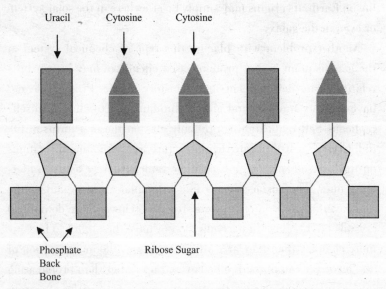

Figure 6.1 The structure of a RNA molecule showing the three nucleic acid bases that specify the addition of alanine to a protein chain.

up in it. It is divided between forty-eight separate strands and the total consists of several billion bases. This sequence of bases specify the exact sequence of the amino acids in the thirty thousand proteins that conspire to make a human being's biochemistry. The proteins themselves are also long strings but they fold into complex three dimensional shapes, after they are produced, in order to perform their chemical functions. This chemical complexity is at the heart of all of the beautiful life that is the highest biological achievement (in our solar system) of this improbably complex universe.

DNA and RNA molecules reproduce in a process that is similar to the growth of a crystal. A single strand of either can serve as a template on to which a second strand can grow. Similarly, RNA and DNA can grow on each other. DNA growth is shown in Figure 6.2. The four DNA bases mutually associate in pairs that tend to stick together (A to T and C to G) and are thus called complementary. During the first step in reproduction, the strand that is produced is not at all like the parent template because its sequence consists of the complementary bases to the original. When this complementary strand becomes the template for the second step, a reproduction of the original strand results.

A--A--C--C--T--G--A--C--G--A--T--G--C--C--A--G--T--G <--Template DNA
T--T--G--G--A--C--T--G--C--T--A   <-- Growing
                                                    Complementary
                                                    DNA
                   C is the next to attach, followed by G

Figure 6.2 The growth of a "complementary" DNA molecule using its "complement" as a template. The complement of A is T and the complement of C is G. In RNA U replaces T.

In a human, the genetic message is stored in the form of DNA molecules. In order to produce a protein, the appropriate segment of the DNA molecule is used as a template to copy a complementary segment of RNA called messenger RNA. It is then used as a template to produce the appropriate protein. Segments of three bases form one of twenty possible three letter words. Each word specifies which of the twenty kinds of amino acids to attach to the growing protein chain next. Some of these chains have tens of thousands of amino acid links. In some cases a single mistake in one of the original bases on the DNA chain can be fatal. Cancer or genetic disease might result. In less than one percent of the cases such mutations are advantageous and become the way of the future. Only two percent of our DNA has changed by mutation in the seven million years that separates us from the chimpanzee.

All cellular life has DNA as the storage mechanism for the genetic code. However it is very likely that this was not the case with the earliest forms of life. Evidence is mounting for the hypothesis that the earliest life forms used the RNA molecule for both genetic storage and for the chemically active functions that are mainly carried out by proteins in modern organisms[6].

In order for RNA to be a good candidate for the Adam and Eve of all life, it should be able to reproduce in an environment that offers little more than the ribosugars and bases that it is made from [7]. That these ribosugars can arise spontaneously, has been demonstrated in Miller like experiments that simulate evaporating tidal pools. However, modern RNA virus must reproduce in the presence of a complex protein that catalyzes (helps) the reaction. The need for this protein makes them less than viable as candidates for the origin of life because this protein did not exist at the start. Recently, progress has been made on this issue by demonstrating that simple and short RNA chain molecules can reproduce directly in the presence of lead and zinc ions [8].

The hypothesis that a strand of RNA was the first form of life has also been gaining support from discoveries that RNA can engage in catalytic functions in the cell. These functions were previously thought to be the exclusive domain of proteins. One example of this is the discovery that messenger RNA molecules can trim out unused segments and splice themselves back together without the aid of any proteins.

The holy grail of this research is then to find a strand of RNA that can catalyze its own reproduction, gather raw materials from a primitive environment, and synthesize the ribosugars and bases that it is made from. Mother Earth provides the cycles of temperature and chemical environment, in hot springs and tidal pools, that may have enabled RNA to fulfill all of these diverse functions. These functions would be kept together not by a cell wall but by virtue of being tied to the same ribosugar back bone. That these functions could be spontaneously spliced together from separate constituents has been demonstrated in a test tube. In a recent experiment in which trillions of different strands of RNA were randomly generated in a test tube, about sixty were found to have the ability to splice themselves together. Alternately, a tidal pool or hot springs with diverse strands of RNA can be thought of as a single organism that is based on a symbiotic relationship between all of the molecules of muck that it contains.

One major deficiency that RNA has as the original life form is that it cannot reproduce in a mixture of left- and right-handed base molecules. The prebiotic environment most likely consisted of such a mixture in which half of the molecules were mirror images of the other half. However, a close relative of RNA that is called PNA (peptide nucleic acid) can reproduce in such a mixture. In PNA the sugar/phosphate groups are replaced by proteins. At this time PNA is shaping up as the prime candidate for the "Adam and Eve" role.

The idea that free-floating RNA molecules were the first form of life is almost like saying that "the chicken came first, followed by the egg". At some point these RNA molecules would have to make or commandeer cell-like spheres in order to look like the early fossilized life that we have evidence of. Once RNA molecules had moved into such an enclosure it would have been much easier to keep the complete complex mix of life-sustaining molecules in close association. They could then work together to sustain all of the necessary biological functions. The high concentrations that can occur within a cell wall greatly accelerate the rates of all of the reactions. As the initial primordial soup, that this life form would have fed upon, was used up by greedy microbes in the Earth's first Malthusian ecological crisis, it would become more advantageous to store the raw material for life within an enclosure. Food and reaction products would then benefit only the enclosed strand of RNA.

An alternate view of the origin of life is based on the idea that the "egg", or cellular container, came first. This idea comes from the observation that cell-like spheres of amphiphile organic molecule form spontaneously [9]. These molecules are similar to laundry detergent. They are long hydrocarbon chains with a water-attracting end and a water-repelling end. These molecules tend to form membranes that curl up into a sphere with the water-repelling surface on the inside and the water-attracting one on the outside. (These are among the types of molecules that have been found in meteorites.) In Miller-type experiments, involving hot lava surfaces, such spheres arise spontaneously. Water-repelling organic molecules from a primordial soup would tend to be enclosed by such spheres [6]. A primitive RNA virus would find such a sphere to be an excellent place to have a feast. If this scenario is correct, the "egg" and the "chicken" arose separately and then formed a symbiotic relationship. Again, the synthesis of diverse forms to make more complex entities, is at the center of many of the great creative events in the history of life. Later this will be seen to operate in the sphere of collective mind as well as in the biosphere.

The hypotheses described above for the origin of life are by no means the only contenders. The surface of pyrite (fools gold) has been proposed as a place where negatively charged organic molecules would naturally attach due to its positive charge [10]. In the resulting community of molecules, many interesting reactions could occur. The impact of giant asteroids has been proposed as the source of the methane that is needed for a Miller-type reaction [11]. Many other speculations abound.

All of these approaches probably owe their existence to an initial proposal by Charles Darwin in 1871 that "warm little ponds" may have been the point of life's origin. In recent Miller-type experiments, it has been shown that all of the four RNA bases (building blocks) can be formed in environments like tidal pools. In such places a mixture of organic chemicals could concentrate by evaporation. The need to concentrate in a community is related to the fact that complex chemical reactions occur much faster when the reactants are in high concentration. There is a common theme to all of these genesis theories. They all propose mechanisms for the concentration of organic material. Examples are: fatty acid spheres (amphiphile), RNA chains, warm evaporating ponds, the surface of pyrite, the surface of clay particles, the surface of evaporating comets, oceanic surface foam, etc.

Just how PNA, RNA, or DNA got together with fatty acids to make the first primitive cells may remain the subject of speculation forever. There is no fossil record of RNA-based life. However the fact that modern RNA uses much more of the C and G bases (vs. A and U) hints at the ancestral roll of RNA [12]. In RNA replication C and G have a much lower rate of error so they would be preferred by natural selection.

As pointed out earlier, the formation of cellular life happened very soon after the last time the surface of the Earth was completely melted by the heat from large meteor showers. In search of a meal, some primitive bacteria developed the ability to convert the iron dissolved in the ocean into pyrite (fools gold). This occurred as early as 3.4 billion years ago [1]. Others evolved the ability to use the chemical energy sources that are generated in hot springs like those in Yellow Stone National Park and at volcanically active regions in the ocean. The most primitive forms of bacteria can be found at these locations. It is very likely that these bacteria are the most primitive forms of cellular life. In fact, life may have originally evolved in hot springs that may have gone many miles into the crust of the Earth. Many of the most basic chemical reactions that support our metabolism occur spontaneously in these deep hot environments [13]. At great depth, primitive life would be relatively well protected from the large asteroid impacts that would have sterilized the surface for the first billion years of Earth's history. Today's hot springs at the bottom of the ocean support small ecosystems that include clams, crabs and giant tube worms that all live off of the primitive bacteria supported by the hydrogen sulfide in the hot water. These bacteria have the ability to live in environments that are hotter than boiling water at the surface. At the depths where they live the boiling point of water is elevated as in a pressure cooker. Because most of our geothermal energy is ultimately derived from radioactive decay the existence of natural radioactivity may fit into the evolution of life as a primary energy source.

Though our opportunistic microbial ancestors were able to make a living off of new forms of geochemical energy, they also were ultimately limited by Malthusian problems. The ultimate solution was found by blue-green algae (cyanobacteria) which evolved the ability to harness the energy of sunlight with photosynthesis. They were certainly around here 2.8 billion years ago and may be as old as 3.5 billion years [1]. In photosynthesis, carbon dioxide from volcanic activity and water (also indebted to the volcanoes) is combined to make sugar and free oxygen. The sugar is then used to make cellular structures such as the DNA backbone and as a source of stored energy. The energy stored in the sugar can be released by the process called fermentation. This process converts sugar into alcohol.

With the emergence of photosynthesis, the overall potential for food increased ten thousand fold; there is ten thousand times more energy coming down to us from the sun than there is heat rising from within the Earth. The larger ecosystem that photosynthesis made possible was crucial to our development. Therefore, this development was yet another biological stepping-stone to the kind of mental life that has a shot at eternity.

This solar solution to the energy problem resulted in yet another Malthusian disaster within 1.5 billion years. About two billion years ago the oxygen that had been released by photosynthesis built up to a level of 0.2% of the atmosphere (it is 20% today). This was a major disaster, yet an opportunity for the life forms of the day. For two billion years life had evolved in a reducing atmosphere (e.g. methane). It was the chemical opposite of the oxidizing atmosphere that we have today. They were not adapted to the destructive effects of oxygen and the biologically destructive free radicals that it generates (many vitamins, such as A, C, and E, neutralize free radicals). Various organisms tried to adapt in various ways. The descendents of those that did not adapt can still be found in deep muds today where the oxygen concentration is below 0.2%. The botulism bacteria can be found in cans that were improperly sterilized. It cannot survive if the container is not sealed. These bacteria failed to adapt to the new environment and so they can only be found in small niches that resemble the old environment.

Some bacteria evolved the ability to combine the oxygen with the iron that was dissolved in the ocean and get energy out of the process. These bacteria are responsible for many of the large iron ore deposits that we use to make our cars. They did well for a while but eventually all of the iron in the ocean was used up and so they did not become the way of the future. They can still be found on the surface of iron objects busily turning them into rust (iron oxide ores).

Other bacteria developed a very effective way of dealing with this major ecocrisis. They evolved the ability to use the oxygen to "oxidize" (burn) sugars to get energy instead of fermenting them. Not only did they get rid of the poisonous waste product (oxygen) but they got a tremendous energy benefit out of it as well. The oxidation of sugars derives ten times more energy than its fermentation. This energy bonanza was soon used to fuel movement as well as metabolic processes.

The particular organisms that are our ancestors, and the ancestors of all higher life forms as well, adapted to the oxygen crisis in a particularly novel way. Instead of changing their DNA, they took into their interior the bacteria that learned how to oxidize. The DNA of these Eukaryotic cells is now housed in the cellular nucleus while the descendants of the oxidizing bacteria float around in the cytoplasm and are known as mitochondria [14]. The mitochondria have their own DNA, and are the power house for modern cells. The symbiotic relationship between mitochondria and these hospitable cells known as eukaryotes [15] is one of the great success stories of natural history. Again, it illustrates the power of community and synthesis. The chloroplasts that plants use in photosynthesis are also the descendants of photosynthetic bacteria taken in by ancient eukaryotes.

The creation of a community of organisms within eukaryotic cells was not all sweetness and light however. The problem centers around what amounted to sexually motivated mortal combat among mitochondria. To see how this problem developed we have to return to the time when DNA-, RNA-, or PNA-based genes first moved into spheres of fatty acids and formed complex cells. Before these genes acquired their new homes it was relatively easy for them to mix genetic material in a primitive version of sexuality. This vastly increased the rate of evolution by allowing separately evolved traits to get together in a more adaptive organism. This ability was retained in modified form even after these genes developed a more reclusive lifestyle inside of cell walls. Primitive bacteria still got together on occasions, opened a common orifice in their cell walls and exchanged genetic material. This can be observed today under a microscope. The exchange of fragments of DNA, called plasmids, is responsible for propagating antibiotic-resistant traits from bacteria that have been adapted to antibiotics to others that have not. Multiple drug-resistant bacteria can arise because two different strains exposed to two different drugs can exchange resistant DNA to their mutual advantage.

A second mechanism for mixing genetic traits is based on the infection of cells by a virus. When the DNA or RNA of a virus infects a cell it frequently combines its genetic material with that of the host cell. The viral genes and those of the host become linked. In some cases, the new copies of the virus can incorporate some of the genetic material of the host. These host genes can then be transferred to a second host in the next generation of infection. (This mechanism of cross-fertilization is in fact exploited in many modern attempts at gene therapy.)

The process of exchanging genetic material between single-celled organisms became more complicated in eukaryotes. If mitochondria or chloroplasts were exchanged at the same time, it probably led to internal warfare between various versions of these somewhat self-contained organisms. The mitochondria that had the ability to kill other mitochondria would become dominant in the subpopulation of these organisms. This would be good for that particularly aggressive version of mitochondria, but it would be bad or even fatal for the eukaryotic mini-ecosystem as a whole. Therefore it became advantageous for eukaryotes to engage in a somewhat modified version of the cell fusion process. In this new "sexual" process one cell (the male) would donate genetic material to a second cell (the female). Only the genetic material from the nucleus (not from the mitochondria) would be involved in this sex act. The female would retain her mutually compatible and non-combative population of mitochondria. Therefore competition between mitochondria was mitigated by the invention of separate sexes. In many organisms (fungi in particular) *more* than two sexes are involved in a complex chain of DNA donations.

The evolution of this new version of "safe sex" was another of the major biological stepping-stones on the path to intelligent eukaryotes like ourselves. Before this division of the sexes, their evolution had been progressing slowly for a billion years, but after this development the pace quickened and the plot thickened. The advantage of being able to combine traits through the sexual fusing of genetic material from different organisms enabled the subsequent biological revolutions that led to our present world. Later we will explore how the free exchange of ideas in communities of mind is the foundation for revolutions in thought and human behavior.

The trail of biological stepping-stones is much longer than the chemical and stellar trail. Only some of the major developments prior to the dawn of multicellular life have been described here. To get this far first required the spontaneous generation of self-replicating and catalytic molecules out of a complex primordial chemical medium. These molecules acquired homes within spheres of fatty acid membranes. They evolved the ability to produce useful proteins. They developed RNA- or DNA-based genetics. They evolved photosynthesis, increasing their food base ten thousand fold. Some learned to oxidize sugars when the environmental concentration of oxygen became too toxic. Others learned to live symbiotically with oxidizing organisms inside them as mitochondria. Male and female sexes evolved and fortunately created a mechanism for genetic exchange without pathogenic mitochondrial competition. All of this took several billion years, but to go from these simple single-celled organisms to us required another billion years and many more stepping-stones.

Chapter 7
Bio Big Bang

A little over a billion years ago, complex organisms began to evolve as a result of a sexual revolution among some eukaryotic organisms. Within a few hundred million years, the first multicellular organisms arose, simple sea weed type plants. Sheer size gave them an advantage against predation by smaller, single celled grazers. Soon grazing animals like simple worms and jellyfish also evolved. The jellyfish had a mechanism for the coordination of movement that was reminiscent of the wave phenomenon that occurs among the fans at a baseball game. Rhythmic movements of the cells propagate across the surface when the motion of a cell stimulates its neighbors to move. Something similar to the propagation of nerve signals also occurs when touch stimulates poison packed cells to release their barbs and at the same time stimulate their neighbors to also release theirs.

One type of organism that supported a complex wiring scheme, and became the way of the future, was the lowly worm. The basic body plan of the worm is a double-walled tube. The inner tube is the digestive track. Initially the propagation of rhythmic contractions could move food down the digestive tract and produce locomotion. Later, a primitive nervous system and brain developed to coordinate these movements. As evolution progressed, greater differentiation of internal organs and body parts led to worms with a triple-wall anatomy with only about one thousand cells. These cells were differentiated into most of the organs of modern creatures like our-selves (trillions of cells).

The worm had a very primitive brain that could actually learn. The slightly more advanced sea slug will initially respond to a touch of its tail by engaging in a flight reaction. It will wiggle violently in an effort to get away. However, repeated stimulation of its tail will desensitize it. This is a very rudimentary form of learning. A long way (half a billion years of evolution) from what could be described as self aware and self reflective, but it was a start.

The evolution of single cell eukaryotes into complex worms took half a billion years. As these worms became better grazers, and even predators, their prey (both algae and animals) developed hard exte-rior skeletons for protection. Soon after the development of these skeletons and segmented bodies, half a billion years ago the greatest spurt of evolution of all time occurred, the Cambrian Explosion. Recent work on sedimentary layers in Siberia [1] indicates that this epoch occurred in less than ten million years. The Cambrian explo-sion, described as the Big Bang of Biology, left the first fossil record of all of the major phyla of animals.

However, genetic studies indicate that the divergence of the various phyla occurred a half billion years before this explosion of diversity in the fossil record. Darwin was puzzled by how so many diverse, complex creatures appeared so suddenly in the fossil record. He even speculated that this was due to the fact that soft bodies are not preserved in the fossil record. A great deal of evolution is still invisible to us today.

There are many theories of the Cambrian explosion. Some believe that the rise of predatory creatures created a horse race between the evolution of predators and that of their prey. The most obvious indi-cation is the sudden existence of hard protective shells or exterior skeletons. Another indication is the presence of holes in trilobite fossils made by a predatory animal with hard mouth parts. Many creatures that evolved at that time seem to be carnivorous.

A second theory of the origin of this burst of creativity proposes that the crust of the Earth slipped on the underlying mantle, changing the latitude of *everything* in a short period of time, by geological standards. According to this idea, every creature was suddenly mismatched to a new climate and those that survived were those that could adapt rapidly. This could be what happens if people suddenly change the climate by emitting excessive amounts of greenhouse gasses like carbon dioxide and methane. The difference, however, would be that creatures that were not weed-like would go extinct. This is because a human-induced change would occur so rapidly that there would not be enough time for evolutionary adaptation. Only tough, well-distributed creatures like rats, sea gulls, and people would survive the kind of very rapid shift in the environment that results from human activities.

Yet another theory of the Cambrian explosion was that it was caused by the accumulation of oxygen in the atmosphere [1]. According to this approach, large multi-cellular organisms became more feasible because it was easier to keep their tissues oxygenated in an oxygen rich environment. This theory, however, does not account for the suddenness of this event.

Very recently yet another possible explanation has become apparent. There is substantial evidence that a series of extreme ice ages occurred in the few hundred million years that preceded the Cambrian [2]. The average temperature at the equator may have swung between −100F and +150F in cycles that lasted fifty million years. The entire surface may have been covered with a sheet of ice many miles thick. The only niches where life could survived may have been around hot springs energized by the volcanic heat of the Earth. In such an environment creatures would have done better if they had a tough body.

Whatever caused this burst of biological creativity was another stepping-stone in our evolution. Our ancestor from this period was a filter-feeding mollusk. As a juvenile it was a free-swimming, fish-like creature that sought out its prey, but as it matured it settled down, attached itself to a rock, and became a sedentary filter feeder. Later it evolved to retain its fish-like, juvenile form into adulthood.

This fish-like creature first evolved into the Carpoidea family (Carp). The variety that developed a spinal cord that ran its full length became the ancestor of vertebrates. Initially, all of the hard body parts were made out of cartilage. (Except for their teeth, sharks are still this way.) Millions of years later, fish evolved the ability to harden their cartilage with calcium deposits to create bone. (Plastic surgeons can turn cartilage into custom made bone today by molding it to a desired shape and treating it with a growth hormone, extracted from pig bone, that causes it to turn into bone.)

Bone was a great material for structural support and for protection. If it hadn't been developed, then the technical representative of life on Earth might of had to evolve out of termites or Fiddler crabs. They developed an external skeleton while we have an internal one. Either is required for land-based creatures. But ours is better adapted to large bodies, because our skeleton does not have to be replaced with every major episode of growth.

About a third of a billion years ago, a fish [3] that lived in tidal swamps (a kind of lungfish) gradually adapted its gills to extracting oxygen directly from the air instead of exclusively from water. There were several possible motivations (selection forces) for this evolutionary development. First the pools in which these pre-amphibians lived dried up periodically. This caused two problems. The obvious one is that in the absence of water to suck through its gills, it would either die of asphyxiation or evolve air tolerant gills, and ultimately lungs. A second problem was that they were smaller than many of the giants that lived in their pools. If the pool merely shrunk, without drying up, then they were forced into close quarters. The ability to get out of the water temporarily in order to escape a banquet in which they were the main course, was a great advantage. A third benefit of leaving the water was the availability of plentiful insects on the land to eat. Here they had no competition from anything that was their size.

Along with their ability to breath air, they also evolved stiff pectoral fins that helped them wiggle their way along slimy mud flats. Such fish can still be found in tidal pools today. (The walking catfish of Florida is a refugee from Africa having been imported for aquariums. It has been known to attack small pets at great distances from the water, often over a bowl of pet food. The walking catfish can cover long distances in search of its next aquatic home, and can therefore find a new home when its old one dries up and becomes inhospitable.)

Over time, such fish evolved into amphibians, which could live out of water for extended periods of time, and whose only absolute need for immersion was for reproduction. They had to lay their eggs in water and the juveniles required it when young. Freed of the tidal pools, they could wander inland to find new habitats. The amphibians spread across the surface of the land to wherever pools could be found during the breeding season. Pools that dried up for part of the year were perfect because fish that ate eggs and larvae would not be found in these temporary "vernal" pools.

Giant amphibians soon became the dominant large creatures on Earth. However, within seventy million years of their emergence from the water, some of their kind evolved the ability to encase their eggs in a leathery capsule. This kind of egg could hatch on land without any water. This creature was the ancestor of the reptiles. Their total freedom from pools of water allowed them to occupy more niches. It also enabled them to hide their eggs better. When they hatched, the young were not crowded into limited pools where they were sitting ducks for predation.

All of these advantages gave the reptiles an edge over the amphibians. However, this edge was not decisive at first. A quarter billion years ago, all forms of advanced life faced their greatest challenge

Table 7.1 The big five killer events of the last half billion years. This is taken from Macolm W. Browne, The New York Times , Dec 15, 1992, page C1.

Event	Date in millions of years	Marine Families Lost	Marine Genie Lost	Big Losers
Final Ordivician	435	24%	47%	Trilobites, Cephalopods, Crinoids
Late Devonian	357	22%	52%	Coral, Brachiopods, Pacoderms, Trilobites
Final Permian	250	50%	78%	Trilobites, Bryozoan, Crinoids, Brachiopods, Forminifera, Ammonoids, Giant Amphiba, Reptiles, Giant Horsetails, Club Mosses, Insects
Late Triassic	198	24%	50%	Cephalopods, Reptiles, Gastropods, Bivalves, Brachiopods, Conodonts
Final Cretaceous	65	16%	42%	Dinosaurs, Ammonites

of all time: the Permian extinction. At this time 96% of all species became extinct. By comparison, all of the other four major extinction events were kids play. These events are listed in Table 7.1 below. Even the incredibly tough insects suffered their only major extinction event in their entire four hundred million year history.

As mentioned before, the source of this great extinction event may have been in the Earth or in the sky. One hypothesis rests on evidence that there was a single land mass at the time, called Pangea. The last time that the Earth had assembled all of her continents into one big one was before Biology's Big Bang a half a billion years earlier. At both times, the vertical circulation of the ocean effectively ceased [4]. The organic material that sunk in the deep sea depleted the atmosphere of carbon dioxide while it poisoned the stagnant bottom waters with a great excess of this dissolved gas. This reduction in atmospheric carbon dioxide reduced greenhouse warming and a great ice age began.

Enough ice piled up on the land to reduce the depth of the ocean by five hundred feet. The edge of the present continental shelf (where the ocean floor slopes down steeply) is where the coast was at the time. The absence of large bodies of shallow water robbed many species of their habitats.

But the worst was yet to come. When the continental glacier reached the sea it chilled the surface waters at its edge, and the sinking of this cold water restarted the circulation of the ocean. The concentration of carbon dioxide and hydrogen sulfide in the bottom waters had reach a level poisonous to most surface dwellers. When these poisonous deep waters rose to the surface, many creatures became extinct. The Trilobites, which had been one of the most successful species since the Cambrian explosion, disappeared. Many other Cambrian success stories failed at this time as well.

Another theory that attempts to account for this disaster rests on the exact coincidence (to within a million years, which is the experimental error) between this extinction event and the emergence of a gigantic flood of molten rock in Siberia [5]. This blob of molten rock had risen from the boundary between the molten core and the plastic mantle at a depth of two thousand miles. It released two million cubic kilometers of lava along with huge amounts of associated sulfur dioxide and ash. The recent eruption of Mount St. Helen involved only one cubic kilometer of lava. Ultimately the flow was a thousand miles in diameter. (It is presently the source of much of the precious mineral wealth of Russia.) The exact mechanism by which this volcanic event might have lead to mass extinction is not known. In all likelihood, its effect on land animals was far worse than on animals in the sea.

A third scenario to explain the Permian extinction is that it may have been due to the impact of a huge meteor [6]. Eventually we will know which (if any) of these hypotheses tells the story as it was. It may have in fact been a one, two, three punch that did it.

Whichever way it was, the bottom line is that crisis produced opportunity. In the aftermath of this greatest of extinctions, the survivors must of had a hay day as new biological niches opened and animals evolved to exploit them. The reptiles diversified and flourished while the amphibians lost ground. After twenty million years, a branch of the reptile family would evolve into one of the most successful creatures of all time: the dinosaurs. They were distinguished from their ancestors by hips that improved locomotion and other features. Over a period of two hundred million years they would dominate the land in size, number, diversity, and distribution. Some dinosaurs evolved into the birds, so some of their descendents are still here today—the hundred seventy million years of their life and death struggles did not terminate in total extinction.

Going back to the Permian extinction, other reptiles began evolving the ability to keep their bodies warm. Later they evolved the ability to nurse their young. These were the mammals. Though the ancestral reptile was fairly sizable (i.e. ten feet), the niche that the mammals occupied caused them to shrink to squirrel sized creatures. They lived in the trees or in holes. At night, when the dinosaurs were immobilized from the loss of body heat, the mammals became active.

The scene might be more or less the same today if it were not for one of the greatest meteor impacts, anywhere in the inner Solar System, of the last few billion years [7]. The deadly meteor must have been about ten miles in diameter. It crashed into the Yucatan peninsula in Central America sixty five million years ago. It scooped out a crater more than two hundred miles in diameter. A mile high wave in the Gulf of Mexico deposited sea floor sediment one thousand feet up the mountains in Cuba and Haiti. Such sediments can also be found several hundred miles inland in North America (Kansas at the time). A global rain of molten debris set off forest fires worldwide. Every creature on the surface of the Earth that was larger than fifty pounds was soon dead and gone forever.

The exact mechanism by which this calamity killed off the dinosaurs, and many other species, is still the subject of debate. The earliest proposal (by the Walter and Loui Alvarezes) was that the cloud of dust, which blanketed the globe, shut out sunlight for an extended period of time [8]. The Earth then cooled. Continental regions became colder than the worst Siberian winter and the dinosaurs froze to death. Some birds, on the other hand, did well by comparison because they could fly to maritime environments kept above freezing by the residual heat stored in the ocean (it did not freeze over). The living fossil called the Komodo dragon is an aquatic reptile that may have survived for this reason. Similarly, alligators and crocodiles could avoid solidification by staying in the ocean.

A second niche environment in which survival was possible was in holes in the ground. This would explain why the largest land creatures did not survive. Snakes, lizards, and small mammals could ride out the winter below the frost line in deep tunnels. The mammals had the additional advantage of internally generated body heat, so they could remain active.

After the dust cleared (it may have taken years), everything had changed. It was like a global reset button had been pushed. The giant reptiles that had ruled the land for the previous two hundred million years were gone. Every creature that was left was no longer optimally tuned to its environment. Once again disaster provided opportunity. As in the case of the Permian extinction, the earth was wide open to a spurt of evolution. This time it was the turn of the lowly mammals and the birds. For us, that meteor was a major stepping-stone.

Starting from squirrel-like creatures, the mammals branched into many forms, exploiting niches vacated by the dinosaurs. Large herbivores and carnivores evolved within about fifteen million years. One of these returned to the ocean during this period of great biological creativity: the ancestor of whales and dolphins. After ten million years, this originally bear-like creature looked like a seal with rear legs and was still adapted to both land and sea. After another ten million years the legs disappeared and their form was fixed into a fully optimized adaptation to an aquatic environment, and whales and dolphins have changed little since then. The land creatures also evolved into optimal forms. Many of the animals of Africa today are recognizable as descendants of animals that existed forty million years ago. In this period, evolution was slow by comparison to the time period immediately after the impact. These forty million year old ancestors are very different from the tiny mammals that existed at the time of the dinosaurs.

After the initial spurt of evolution, the pace of change slowed as it always does, a process called Punctuated Equilibrium. Darwin initially proposed that species evolve slowly and continuously over vast periods of time, but it now appears that periods of rapid change (in response to new environments) are mixed with periods without much, because once creatures are optimized to their environment there is less pressure for them to evolve further. Therefore, the rate of evolution is tightly coupled to environmental instability. Rapid evolution occurs when massive environmental change occurs infrequently enough for changes in the environment to be adapted to (more than ten million years does it). But these changes must not be so infrequent that stagnation dominates.

Stagnation is life being less than it might become. If it had occurred for too long at any of the major eras in our history something like us might have ended up in the iron age when the sun entered into its death throes. The sun has a clock and so does evolution, and a blissfully stable environment like the womb of the ocean, where whales have found a stable form for the last twenty million years, is also a kind of death trip. Unfortunately, it is now vulnerable to a more highly evolved predator (i.e. us).

Death and destruction are an inherent part of the process of life. It flogs us up the evolutionary tree. Life is a phenomenon that occurs on the razor's edge between the stasis of the rocks and the oblivion of decay. The tension of life and death struggles makes it life. Anything else is ultimately death. If a Creator had no choice but to create in this motif, then so be it. It is hard to imagine how the energy of the Big Bang could have spontaneously transformed into people in any other way.

Chapter 8
Madonna and Child

The idea that humans evolved from "nature, red in tooth and claw" scandalized many Victorians. Darwin's new theory not only contradicted the story in the Bible on a literal level, but it also seemed to fly in the face of the spirit of Christian teachings on brotherly love. How could love exist in a world where competition for survival is the fundamental engine that drives evolution? Where did love come from? Recently, many sociobiologists have racked their brains over this issue. They have come up with the concept of "kinship" to explain how an animal can achieve survival advantage for its genes by helping another animal survive. On the most basic level this is why a parent nurtures their young. This kinship-based altruism generalizes with reduced intensity to siblings aiding each other (on occasion). Genetic relatedness is also one of the principle motivations for communal spirit in primitive societies. These societies are often based on kinship groups. The most basic kinship group is the mother and child. This primary relationship is the well spring of human altruism. It also became the primary source of received knowledge.

When a group gets so large that its members are only distantly related, the kinship motivation for altruistic behavior becomes insignificant. Concepts of alliance for mutual benefit help fill in the gap. A herd has many eyes to detect a predator and the chances are that only one of the many will fall prey. Reciprocal behaviors are another motivation for altruism [1]. Vampire bats will share blood meals with a particular ally, such that both members of the alliance are buffered from death from starvation which occurs if they are not fed every 36 hours. Macaque monkeys who do not announce the presence of a newly discovered food source to the troupe will be picked on more frequently by the other members if they are discovered to be hoarders. Young male baboons will be rewarded by higher social status and commensurate sexual privileges if they risk their lives defending the troop from predators.

All of these mechanisms for altruistic behavior enhanced the survival of the community and the individuals in it. In general, the initial evolution of altruistic behaviors depends on a kinship-based benefit. In large human societies this kinship anchor is often lost, as are a great deal of the altruistic behaviors found in small communities. Reciprocity and social contracts help fill in the void. But in truly mass societies lawlessness becomes a problem. A legal structure to deal with this is a poor last-resort. What is needed are individuals that perceive their relatedness on a psychological level as opposed to the biological level, something that all of the great world religions have attempted to do. In many religions we are seen as "the children of God" and are therefore related. We will see later that a concept of psychological relatedness arises from the sphere of learned behaviors. Who we are, in a behavioral sense, is only partly determined by our genetic inheritance. In large measure, the heart of our psychological being is also formed by learning from other people. This occurs most powerfully with people that we love. These are people who are soul brothers, sisters, children, and parents. Though this type of inheritance does not require biological kinship, it began with it. To see this we have to back up a third of a billion years in evolution.

Reptiles, amphibians, and fish are very tough, self-sufficient animals. Most of their social interactions involve herding for protection from carnivores and brief sexual encounters for reproduction. Once eggs are fertilized, the relationship ends. A huge multitude of eggs is usually deposited in some kind of nest, a strategy of quantity instead of quality. The offspring will hatch simultaneously and obtain some safety from the fact that many will get away while some are eaten. If a nest is discovered, perhaps the predator will eat its fill before the last egg is gone. The investment of effort in building a warm hidden nest is rolled over a large number of eggs so the mother, and sometimes the father, usually works at building a good one.

The nest is usually abandoned once the last egg is laid. The parents will have nothing further to do with the young unless it is to make a small meal out of a slow one. In all likelihood the adult is unrelated to its prey, so there is no evolutionary expedient in protecting the young as opposed to eating them. There are notable exceptions to this situation however. The sea horse male receives eggs from the female and broods them in a pouch. When they are ready he gives birth to fully developed fry. The male of the Siamese fighting fish builds a nest of air bubbles in swamp slime and places the eggs from the female in the nest. He then drives off the female and guards the nest from predators that include the cannibalistic mother and other males. He will fight fiercely in defense of the nest. When the eggs hatch he will baby-sit the nest. If any of the fry stray from the nest, he will suck them into his mouth and expel them back into the nest. Whether or not any of these complex behaviors are learned or are purely instinctual might be debated. In all likelihood they are purely a matter of hard-wired instincts. A young male that strayed from the nest soon after hatching would be fully capable of performing all of the necessary tasks.

The female Nile crocodile also protects her nest. When the eggs hatch she will shelter the young in her gigantic mouth. When they are bigger she will keep them close. Babies can often be seen covering a female's back as she glides through the dangerous waters. Again, these behaviors do not require that the young learn from the mother. Instinct alone is sufficient.

During the reign of the dinosaurs, some duck-billed herbivorous dinosaurs (Maiasaursa) built nests in communities for mutual protection. There is even evidence that the mother guarded the nest and fed the young after they hatched. There is also some controversial evidence that some dinosaurs were pseudo-warm-blooded. Warm-blooded animals have the tremendous advantage that they can remain active even when the weather turns cold. A warm-blooded animal can prey on the nest or body of a cold-blooded animal when it is too cold to function.

There is a large economic penalty for being warm-blooded however. Body heat is maintained by burning energy even when the animal is inactive, requiring extra food. The farming of cold-blooded fish is much more efficient than that of warm-blooded cattle. Two pounds of grain will make more than one pound of fish. With cattle and other mammals, the ratio is ten to one. The tiny shrew needs to eat almost its weight every day to stay alive. The even smaller humming bird must eat twice its body weight per day in sugar-rich nectar. Because they are small they have a lot of surface area by comparison to their weight. This is also true of newly hatched or born mammals and birds, who have prodigious food needs both for rapid growth and to maintain high body temperatures. The parents must help them find food immediately or feed them directly.

In the case of mammals, milk is provided from the mammary gland. It is at the breast that the story of maternal love took on new meaning for our kind. The close bond between mother and child that results from breast feeding is the foundation for the evolution of many behaviors that can be broadly categorize as love. The nurturing of an animal by another is the foundation for the complex cooperative behaviors that have brought us to our present level of cultural, psychological, and scientific development. The need to feed is followed by the need to be taught how to feed one's self.

In the teaching of behaviors lies the beginning of what Pierre Teilhard de Chardin calls the Noosphere [2]. Concisely stated, the Noosphere is an ecosystem of mental entities that resides in the collection of human minds. It is the psychological analog of the biosphere. The passing of ideas, behaviors, and traits through communication and learning is the foundation of this new sphere of reality. Other words can be used to describe this new sphere of existence: culture, collective psyche, learned behavior, tradition, and human spirit. In this new sphere of existence, psychic essence is encoded in the synaptic connections of the brain instead of in the DNA base sequences. Whereas DNA dodges mortality through sexual communication, this new form of psychic entity does something similar through communication between individuals. Our evolution into this new way of existing and evolving is the most important stepping-stone to our kind of self-reflective life.

Teilhard de Chardin places the beginning of the Noosphere at the start of the Pleistocene geological era three million years ago. At this time apes were beginning to noticeably evolve a human-like shape. I would place its beginning very much further back in time and not limit it to the human species—back to the time when mammals in our line first began learning from their mothers and their siblings. At first the lessons were simple and constituted a tiny fraction of the behavioral repertoire that an animal needed for survival. Subsequent evolution has brought us to the point where learned behaviors are all-important to survival and yet instincts fulfill simple rolls such as eating when hungry. (Even instincts as fundamental as this can be overridden by the teachings of *Vogue* magazine on the nature of the ideal female body. Anorexic behavior is a uniquely modern phenomenon driven by cultural imperatives.)

The human collective psyche has evolved to such an unprecedented level that it is easy to see it as unique. It is unique in its quantitative level of development but not in its qualitative existence. Other species have their own collectivity of mind. Whales sing songs that gradually evolve [3]. Some species of birds do the same. Though beavers are born with a propensity to dam streams with mud and sticks, they begin this behavior late in life and learn many refinements from their parents. We will see later that elephants depend a great deal on learned behaviors. Therefore people do not have a monopoly on the kind of collective learned behaviors that constitute a Noosphere.

Not only do other animals have their own Noospheres, but in a weak sense we and they are joined in an even larger sphere. The fact that many Paleolithic people (precivilization) used animals as "Totems" (spiritual role models) suggests that humans have been connected to the psychological world of animals for a long time. Whether or not they learn anything from us is a fascinating question.

The alliance between instinct and learned behavior goes very far back in the history of the animal world. In a very primitive sense even plants learn. In response to infestation by disease or insects they will initiate the production of chemicals that fight such parasites. Our immune systems "learn" to fight specific infections. Primitive worm-like sea slugs learn to turn off their flight reaction after their tail has been tickled enough times by a researcher. This latter form of learning is not an alteration in a chemical system so much as a change in structure and chemistry in a nervous system. It is therefore more akin to what we know as learning. This kind of learning became basic to the nervous systems of all animals more than half a billion years ago. It was required before a noosphere could evolve but it is still a far cry from it. It is experiential learning from the environment, as opposed to from other minds. To the extent that other minds are part of the environment, the distinction is artificial.

The origin of mammals can be traced back to the time of the great Permian extinction a quarter billion years ago as the demise of 96% of all species occasioned a new flowering of biological creativity. The reptiles that managed to survive this calamity diversified into many new forms. Some evolved into dinosaurs and later into birds, while others perfected their reptilian forms. A particular group of reptiles, known as therapsids (not therapists quite yet), began to develop mechanisms for controlling body heat. The structure of their teeth suggests a greater amount of chewing than was characteristic of other reptiles, which may indicate the need to process more food to maintain body heat. Also, their nasal passages appear to have been capable of heating and moistening air as in modern mammals. Air passages being routed around the mouth would have allowed for breathing while chewing. The structure of the rib cage suggests a more muscular diaphragm for enhanced breathing. All of these innovations point towards a higher metabolic rate, and in turn suggests warm-bloodedness.

One of the more primitive therapsids was called a dimetrodon. It could grow to a length of 12 feet. It looked like an alligator with a large sail-like fin running down its spine. The fin was probably used to regulate body temperature. When it crawled out into the bright morning sun, it would turn its side to the rising sun. Blood, pumped through the fin, would warm quickly. This gave the Dimetrodon a jump on the other reptiles of its day, as it could be warm and mobile long before they could. Other reptiles had used solar heating to control body temperature but this solar panel went an important step beyond: the fin could also be used to expel excess body heat, cooling its blood with breezes.

Fifty million years after the therapsids we find reptiles so mammal-like in appearance that experts cannot decide which side of the line they are on. These animals, called tritylodonts, had rodent-like skulls, and large front teeth for gnawing, separated by a large gap from the back teeth that were used for chewing. They were probably herbivorous. Notably, they had lost the fin of the dimetrodon and were much smaller. Therefore, they had to rely on internally generated body heat to stay warm. This independence from the heat of the sun would allow them to prowl about at night, when the reptiles were dormant. They soon filled this essentially unoccupied niche.

The first undisputed mammals, the triconodonts, appeared shortly after the tritylodonts. Some were carnivorous while others were insectivores. The largest ones grew to the size of a cat. Perhaps their only living descendants are the monotremes of Australia the duckbill platypus and the spiny anteater. These hairy animals lay eggs, but they carry their young around in a pouch after they have hatched. Yet, because they nurse their young, they are true mammals.

The mammals that were eventually to inherit the Earth were the Pantotheria. Appearing over a hundred million years ago, at the height of the age of dinosaurs, these small shrew-like insectivores gave rise to the marsupials like the kangaroo and the placentals like ourselves. The Pantotheria were distinguished from the triconodonts by their lack of egg laying. Instead of depositing their fertilized eggs in a vulnerable nest, the mothers retained them in their own bodies. This was an advance in the intimacy between mother and child: now the young developed in a well protected and mobile womb until they could reach a level of development that allowed them to suckle.

The marsupial young emerge in an extremely immature state. The joey of the kangaroo, for example, looks like little more than a worm when it is born. It crawls up the abdomen of the mother to the pouch. Here it finds and attaches itself to a teat, capable of little more than suckle. By degrees it matures and begins to make forays outside of the pouch.

In contrast, the placental mammals retained the fetus in the womb to a greater degree of maturity. In the case of many herbivorous mammals, the newborn animal is capable of keeping pace with the herd within hours of birth, and within days it can potentially evade a predator. The key adaptation was a biochemical mechanism in the membrane of the placenta that surrounded the fetus, protecting it from attacks from the mother's immune system, the cells of the fetus being foreign to the mother's body. (Many mothers get morning sickness as their immune system gears up to repel the perceived invasion.) In an interesting twist, this mechanism depends upon the fetus being sufficiently *different* from the mother to trigger the protective mechanism. Thus, many cases of infertility are caused by the father being too immunologically similar to the mother.

In addition to the antibody barrier, the placenta refined its ability to supply oxygen and nutrients to the fetus. The ability to retain the fetus in the womb to a greater level of development gave the placental mammals a significant edge over the marsupials.

Of course, it was not only in the physical protection of the young that placental mammals realized significant competitive advantage—birds, alligators, and ants all care for their young. What led to the dominance of mammals was brain power. Warm blood not only enables muscles to function at peak efficiency, but it also helps neurons do so. A constant body temperature allows nerves to repeat their impulses more accurately, which is essential to memory and learning. The relatively large body sizes of ground dwelling creatures does not limit brain size in the way that flight requires of birds. Mammals have all the brain structures of birds and reptiles, but they also have a huge neocortex, or new brain, the seat of analysis and learned behaviors. The combination of a brain with the ability to learn new behaviors with nurturing parents is, in my view, the foundation for the noosphere. Perhaps this is why the Madonna and Child icon is so important to us (see Figure 8.1). For two hundred million years, any behaviors learned from others were almost exclusively learned from the mother. This new ability opened up a new form of behavioral inheritance. Previously, DNA had to be altered by random mutations within the process of natural selection in order to change a species' behavior. This is a process that takes tens of millions of years and is therefore inefficient in the face of rapidly changing environments such as during the onset of an ice age. On the other hand, learned behaviors can be altered many times in a single generation. By trial and error an animal can discover what works in a particular context, and successes are often (but not always) passed to the next generation. To paraphrase Darwin: It is not the strongest, nor the smartest, that survive—it is the most adaptable.

The tremendous acceleration in the rate of behavioral evolution, resulting from the combination of nurturing (love) and powerful brains brought humanity to where it is today. It has enabled science, art, music, technology, and modern economies—in short, civilization. However, at times this new talent seems to be more of a curse than a blessing. The ebb and flow of destructive behaviors, like aggressive nationalism, drug use, and disregard of life itself, are part of the price we pay. Creatures ruled by instinct have an anchor attached to hundreds of millions of years of evolution. A modern rock groupie's only anchor might be the ravings of a drug-crazed idol.

Though the learning of behaviors among mammals began with the mother-child bond, it soon involved bonds to siblings, surrogate mothers, and other members of the community. So, learning not only propagates down through generations, but also horizontally within a generation. The ability of successful behavior patterns to reproduce in this way is the fundamental phenomenon that makes our collectivity of mind a kind of ecology. Minds linked by this mental analog of biological sexuality are the medium in which the intellectual content of humanity evolves.

Figure 8.1 Madonna and Child [4]

Every individual human personality is a unique combination of biological nature and psychological nurture. Who we are in a behavioral sense, or in other words our soul or character, is inherited through our genes, learning from other people and interactions with the physical environment. How much is "nature" and how much is "nurture" will always be debated, but it is undeniable that humans have a great plasticity of behavior coupled with an equally great dependence on learning.

The entities that have evolved in the human Noosphere seem nearly as varied as the biological entities that have evolved in the biosphere. In addition to the learned aspect of personality there are numerous other entities that exist, propagate, and evolve in the collectivity of human minds [5]. Trivial examples include the rage of swallowing gold fish in the 1930's. Hula hoops and Elvis Presley were fun for a while in the 1950's and the Beetles and psychedelic drugs were in the 60's. More important developments might include the works of Mozart and the teachings of Buddha, Confucius, Moses, Mohammed, Gandhi, and Jesus. These extraordinary people gave birth to many psychic progeny. They have deeply influenced millions of people. They helped shape the souls of future generations with the beauty and love contained within their own souls. This is what mammalian mothers have been doing on a more individual level for two hundred million years. To see this we need to focus on a single species that is different enough from us for its similarities to stand out. The elephant is just one example among many mammals that depends heavily on the transmission of behaviors from the adults to the young. To see their spirit requires that we look deeply into their daily lives.

Chapter 9
Elephant Spirit

The harsh conditions that led to the extinction of the dinosaurs favored warm-blooded animals that nurtured their young—the birds and mammals. Their nurturing went well beyond the provision of food. It included teaching the young how to find food, how to avoid predators, and how to be social. The combination of instincts and learned behaviors in individuals and in groups created the psychic makeup of these modern animals, a totality that I call *spirit*.

Nature and nurture have both contributed to the spirit of these advanced beings, a dichotomy well illustrated by the richness of the spirit of elephants. Depending on the circumstances and the animal, an elephant's behaviors will be dominated by passions or by thoughtful application of experience and culture. It has the largest brain of any land animal. Even though herbivores are not noted for their mental powers, many naturalists consider the elephant to be one of the most intelligent animals. Its sixty year life-span is matched by the length of time that it requires to reach various stages of maturity. As with humans, sexual maturity begins after at least a decade of life. Also, a great deal of learning and experience is required before they reach the full flowering of their being. Male elephants are driven from the family when they are about ten years old to avoid incest, and they spend years alone or in small groups of males. They do not become fully capable of competing for mates until they are forty years old. Females do not become the leaders of their family group until a similar age.

Because of its huge body, the adult elephant is immune to most predators. This size also increases its digestive efficiency up to 50%. This may seem paltry by carnivorous standards but it is quite an achievement given the course vegetation that the elephant consumes. An adult eats over a hundred thousand pounds of food per year. In seasonally bad times it relies on stored fat in its huge body. The major drawback of a large body is lack of speed, for while this is not much of a problem for an adult, it can be fatal to young elephants. We will see that this vulnerability is the root motivation for many of its "family values".

Modern elephants lack the strange shovel shaped lower tusk that their root-eating ancestor, Moeritherium, used to dig for its living, but in bad times they will dig up the roots of grasses with their tusks. After carefully shaking out the dirt, they eat the roots. Young elephants, who have not learned this skill yet, must be content with eating the dung of adults at such times of scarcity.

The elephant's life is built around a family unit led by the oldest cow. While many species depend upon quantity of offspring to overwhelm the grim reaper, the elephant, like ourselves, relies on quality. As with people, a great deal of effort is expended in conceiving and rearing the next generation. A female becomes fertile only in good times and not until several years after the birth of a previous calf, and remains in estrus for only a few days. During this time both she and her family will engage in vocalizations to attract mates. Bull elephants, who may have been waiting most of their lives for this opportunity, respond quickly and come from miles around.

After much signaling and maneuvering among the bulls that might develop into mortal combat, one will establish his dominance. He will typically be large and at least forty years old. He will be at the high point in his annual sexual cycle. In this state of musth he is a totally different animal. He has what could be a fatal overdose of testosterone (four times normal) in his system and aggressiveness to match. Only another bull in musth would be motivated enough and brave enough to risk challenging him. Usually, only his posturing and previously established rank is sufficient to get the other bulls to back off. He will then mate with the fertile cow. If he is old and experienced he may monopolize her until she is no longer fertile. In order to make sure that she has the best mate, the rest of a female's family will engage in a loud mating celebration immediately after copulation is complete. This will attract other bulls to the scene.

If the dominant bull has to fight off a rival, a younger and less dominant bull might succeed in mating with the cow while the dominant one is distracted. Therefore the younger bulls passively hang around and hope for the opportunity. This is one of the few strategies that they can use to successfully mate. They have to learn how to feign a lack of interest when the dominant bull is around, and yet remain close enough to the cow in estrus to exploit any chance that may arise. Any indication that they are seriously interested in the cow will precipitate a violent response from the dominant bull. A fight to the death or a high-speed chase that could go on for miles might result. A less dominant bull in musth will even shut down the external signs of his condition in order to avoid inciting combat. These behaviors are learned from experience with other bulls. They are used to keep what could be fatal instincts (e.g. the sex drive) in check until they can be expressed safely.

A pregnant cow carries its fetus for 22 months. The birth is inevitably difficult because the calf will be almost three hundred pound and three feet tall. If the calf is unusually large (because the sire was large) or if the mother is young (e.g. less than eleven years) and small, the birth could be particularly difficult. A cow's mother, sisters, and cousins attend all births and offer encouragement and advice on stance. When the calf is born, they help it to its feet so that it can reach its mother's breast. If it fails to do this, the mother and some of the family stay with it for as long as is necessary. Then, the mother may attempt to carry it to a location hidden from predators and shaded from the sun. If the infant dies, she remains with it until the smell is an obvious signal that its life is lost. If this happens, it is a great source of sorrow for the entire family, and the mother in particular. Frequently an attempt will be made to bury the dead calf. The sight of it being dismembered by scavengers is repugnant to the entire family and they will often chase off scavengers for a long time.

The happier outcome is more likely. Once the calf is walking, the family will move slowly at first in deference to it. Unlike a young wildebeest or gazelle, the young elephant is incapable of outrunning a predator. If a threat develops, the mature cows will form a line of defense. Having many mature animals in a family group is important in Africa, as predators often form hunting parties—one cow could not defend her calf alone against a pack of persistent hyenas.

Soon after birth, the family will no longer make allowance for the young calf. This is tough love. It must soon keep up with the family and negotiate the obstacles of its environment, scampering over logs, up and down riverbanks, and through swamps. In one part of southern Africa, where salt is scarce, elephants must learn to navigate great distances in pitch-dark caves to find the salt necessary for survival. In all parts of Africa and Asia it must learn to swim through swamps, rivers and lakes with only its trunk above the water like a snorkel for breathing. Its instinctual distress cry is its backup system when stuck in the mud or lost in tall grasses.

The family unit in elephant society provides much more than mutual protection. It is also the environment in which learning occurs. The ultimate source of experience and received wisdom is the matriarch of the family. She must lead her young through difficult dry seasons when food and water are hard to find. At such times it becomes necessary to dig into a dried-up river bed to expose precious hidden water. Smell and knowledge of where and how to dig is one key to the families survival. When danger is present, it is up to her to evaluate the situation and to either lead an organized defense or lead them to safety. When starvation threatens she must draw on her many years of experience and the knowledge that has been passed through many generations to find the tons of forage that the family needs each day. By learning from the matriarch's example, the younger cows acquire the ability to assume the roll when its time comes. Calves learn from all family members, and from peer relationships outside of the family, too.

The dependence of elephants on learned behaviors is well documented by Cynthia Moss in "Echo of the Elephants" [1]. Studying elephants in the Amboseli National Park of Kenya for two decades, she spent a full cycle of the seasons focused on a single family led by a gentle cow whom she named Echo. She writes:

"As well as defense against predators, the family unit is also an important environment for the nurturing of calves. An elephant takes a long time to mature and in the process requires a great deal of attention and teaching. There is continuing debate about which aspects of an animal's behavior are innate and which are learned or acquired. Elephants are doubtlessly born with some instinctive behavior patterns, but it appears that a great deal of what elephants eventually do as adults has to be learned."

This dependence on wisdom and personality, passed through the generations, makes elephants part of a pachyderm Noosphere or culture. Such a culture may not seem like much by human standards but it was equal to our own a few million years ago. In some ways it exceeds ours today—many modern humans could learn a lot about childcare from wild elephants. (But they, like us, have difficulty with artificial environments. The mortality rate for elephants born in zoos exceeds 50%, mainly due to infanticide.) In the wild, the task of raising a calf is not the sole burden of the mother. She can count on the enthusiastic assistance of the immature cows in the family, though she may find it necessary to drive some of them off to avoid overcrowding of the newborn calf. She may pick a particular one to become the calf's special "allomother." The role of an allomother lies somewhere between that of a baby-sitter and a full-fledged foster parent. She keeps close tabs on the calf whenever the mother is occupied in some other task such as feeding or bathing. A fortunate calf has a multitude of allomothers, enhancing its chances of growing to maturity in a hostile environment. If the calf engages in dangerous or unacceptable behavior, they make their disapproval apparent. If the calf playfully charges off into the bush, the allomother(s) will be close behind. Frequently, they engage in play with the calf.

As with other mammals, play is an important learning mechanism. Soon after a calf is born, one of the first things it learns is how to use its trunk. The trunk has forty thousand tendons and muscles that control its movements, and mastering this complexity is no easy feat. So right from the start, the calf plays with its trunk. At first, the calf can do little with it beyond letting it droop down and swinging it about. By sticking it in its mouth and sucking on it, the calf begins to learn some control. Playfully trying to grasp sticks and pick them up is the next step. Ultimately, its trunk becomes its principle tool for feeding. By example, the calf learns to grasp a bunch of grass and kick it with a front foot so that the grass shears off at the roots—just one of the hundred or so different sources of food a calf learns to exploit. It will learn how to fill its trunk with water and squirt it into its mouth for drinking, or over its body for bathing when the water is too shallow to wade. Learning how to give itself a dust bath, with a trunk full of dirt to control parasites, comes later. As it gets older and the trunk gets longer it learns to use it to generate the sounds that elephants use to communicate.

Human languages have evolved hundreds of thousands of words, making it easy for us to discount the significance of animal "languages" that typically involve only twenty-five to fifty sounds and gestures. But it should be kept in mind that the two million years that it took us to acquire this complexity is a recent moment in the history of life. It should also be noted that people typically draw upon only six hundred words to express themselves. The true power of language derives not from the number of words but from the ability to combine them into complex meanings. Numerous experiments with teaching primates to use sign language indicate that many have the innate ability to create complex meanings by combining symbols. A gorilla named Koko [2] even invented "white tiger" to describe a zebra, "finger bracelet" for a ring, and "eye hat" for a Halloween mask. Her abilities and those of other primates with sign language are uncanny.

As with most other mammals and birds, elephants use many sounds, gestures, and smells to communicate. The sounds that elephants use are frequently "infrasonic," at a frequency too low to be heard by humans [3]. These low frequencies can propagate over great distances without being absorbed by trees and other obstacles. In the Alps and the Himalayas, humans use long horns that generate low notes to signal over great distances. The infrasounds that elephants make can be detected by other elephants at a distance of six miles. Things like "where are you?," "I'm here," "I'm in musth," and "she has just mated" have unique sounds and infrasounds. A lovesick refrain, sung by a cow in estrus, may be repeated many times for the better part of an hour. How to make these sounds and what they mean is learned from other elephants. At close range there are many sounds and gestures as well. Greeting signals are obviously very important to elephants because they have so many of them. A particularly exuberant reunion is described by Cynthia Moss [4]. Emily, a female elephant, and her two calves had been separated from Echo's family for a week. After making repeated contact rumbles, Emily finally received a return signal to which she replied in a distinct fashion. She and her calves then made a beeline to Echo and the rest of the family. Echo and Emily greeted each other as follows:

"The two of them raised their heads in the air, clicked their tusks together, entwined their trunks and roared, rumbled and trumpeted while raising and flapping their ears, and spinning and turning, defecating and urinating. All of the others joined in as well making an incredible amount of noise with their loud almost liquid rumbles."

This display went on for ten minutes. It shows the strength of the bonds of love within an elephant family and the numerous signals that elephants use to communicate this love. The use of numerous signals together constitutes a kind of parallel sentence as opposed to the sequential sentences that we use. These bonds of love were further demonstrated a few years later by a reunion that occurred after Emily's poisoning death from browsing at a garbage dump. One day, Echo's family happened to be passing near Emily's bones. Though Emily's remains were down wind and out of sight, the family went straight to her bones. The younger elephants got there first and began to caress the larger bones reverently. Emily's daughter, Eudora, pushed through the crowd and "concentrated on Emily's skull, caressing the smooth cranium and slipping her trunk into the hollows of the skull. Echo was feeling the lower jaw, running her trunk along the teeth—the area used in greeting when elephants place their trunks in each other's mouths."[5]

The mood of this meeting was broken when a one-year-old calf, which had not known Emily, began tossing some of her bones about. Undoubtedly, he would learn what was going on when he grew older and experienced the pain of loss that death brings. Soon after this interruption, the family moved on, but some of the elephants carried off bones in their trunks. This type of behavior is well documented in many other cases. Sometimes the bones are buried. Just what it means to the elephants and why they behave in this way is not known. However, it is clear that love is at the heart of it.

This strong and enduring love between members of a family exists because it has enhanced survival over tens of millions of years. It is absolutely needed for elephants to survive in a complex and at times hostile environment. It is the basis for: alliances for defense against predators, sharing of resources, and sharing of received wisdom. A Darwinist might ask: "where is the competition?" The competition between bulls for mates is obvious. Less obvious is the competition and alliances between families. This becomes important in times of scarce resources. At these times, dominance counts for survival. Rank is frequently asserted when the opportunity arises. The structure of dominance becomes visible when elephants gather into large herds for socializing or migration, forcing contact between families. Families that are friendly with each other will tend to stay near each other. Their calves will frequently play with each other. On occasion a brush with a hostile family will occur. At these times it is up to the matriarch to establish the rank of her family. Often the young calves become pawns in the game. They can be brazenly kidnapped by a hostile matriarch and roughed up a bit by her family [6]. Though the calf is seldom seriously hurt, this behavior lets everyone know who's boss. The calf in particular gets a lesson that it will never forget. It learns the structure of elephant society and where its family fits into it. And it learns a behavior that it will probably repeat when it grows up. It is interesting how both love and aggression have the power to propagate from one generation to the next and across a generation. This open-ended characteristic means that the consequences of either kind of behavior cannot be predicted. Great growth or destruction might ultimately result from behaviors that are intensely loving or aggressive.

This propagation of learning across generations and within generations is the defining characteristic of a Noosphere. When a behavior or idea propagates throughout a population it becomes part of the essence of the animal. It is encoded and preserved in the synaptic connections in the brains of all of the animals in the population. This is analogous to the encryption of biological characteristics in the gene pool that is the collection of all the DNA in all of the animals of a species.

In the case of the learned behaviors, each individual plays out a particular behavioral pattern. The content of the mental pool of behaviors evolves as these behaviors are tried and accepted or rejected as either life enhancing or degrading. The animal usually makes the evaluation on the basis of short-term pleasure or pain.

There is tremendous adaptive power in this new way of evolving behaviors. Sadly, the ultimate factors in adaptation are not short-term feelings but long-term consequences. Emily found out, too late, that eating from the garbage dump was not a good long-term practice. Whether or not she communicated this to her family is not known, but eating at the dump was not taken up by the rest of them. Similarly, people often find that pleasure-giving activities lead to pain in the long term. Drugs, alcohol [7], unsafe sex, and deficit financing are but a few examples. One of the principle functions of the neocortex is to evaluate of likely future consequences, then to impose on the imagination the associated feelings. In this way the decision-making process incorporates a longer range evaluation of well being.

Though we have evolved a Noosphere of unprecedented complexity and power, people should not be too quick to assert their superiority. So far we are a flash-in-the-pan species. Whether or not we have what it takes to match the long evolutionary history of the elephant remains to be seen; the story of the tortoise and the hare comes to mind.

Like ourselves, elephants have evolved over tens of millions of years. They have acquired many traits that we find admirable and some that are not so appealing. The two subspecies of elephants that remain today are the sole survivors on a family tree that had six hundred branches. Their traits were forged by the life and death struggles of countless animals over countless generations. If humans can't see fit to allow them a toe-hold on existence in this century, we will have committed an unspeakable genocide. Our descendents will mourn the loss of the elephant and will consider us to have been greedy barbarians, a possibility that applies to countless other noble creatures.

Chapter 10
Human Spirit

What a human is or should be is a question that six billion people must each deal with in his or her own way. We are not born with the hard-wired behaviors of ants. We have instincts, but, as with elephants, they become highly modified. Most of what we do as adults has to be learned. In large measure we construct our psychic essence, or spirit, from our experience and received wisdom. Innate tendencies are important, but they do not dominate actions as they do with lower animals. This has led to the flowering of human personality, the arts, science, sports, and innumerable other activities that are uniquely human. However, as with the Greek heroes, our greatest virtue is also our greatest weakness. The ability to be almost anything also allows us to become monsters and leaves us adrift with the question of what we should become. Like Icarus, we can now fly too close to the sun.

In the evolution of the human brain, the importance of the neocortex has increased relative to that of the old brain. The new brain is wrapped around the old brain and their activities are intimately coupled. It has evolved from a collection of minor processing areas into a dominant player in perception, analysis, and decision making. The newest parts of the new brain are in the frontal lobes. It is here that associations between diverse perceptions are integrated to form a coherent internal concept of the external world. In relationship to total brain size, our frontal lobes are three times larger than that of a chimpanzee and ten times that of a cat. Our old brain is very similar to that of lower animals like reptiles and birds. It is where emotional reactions are generated and many pleasures are felt. Emotion-based decision making was the old way. On occasions when decisions might lead to long-term disaster, the new brain figures out the problem and makes adjustments.

Recent studies have found that either the old brain or the new brain can initiate motor activity, motor activity being where the rubber hits the road. If a threatening situation develops quickly, the old brain can unilaterally react much faster than if the new brain gets involved, great in emergency situations. Proficiency in sports and combat depends upon this. In more relaxed circumstances, our perceptual inputs are analyzed more carefully by the new brain and the appropriate responses are generated by it, though the emotional response of the old brain is folded into the ultimate decision.

Conflicts between the emotional needs of the old brain and the decision-making criteria of the new brain are a rich source of income for psychiatrists. A resolution in which both are used is a good indication.

The flexibility of the new brain has allowed humans to adapt to every environment found on the planet. Along with the ability to become anything comes the need to learn to become something. Children have a great need to play and to learn. The need is filled by providing them with a rich environment, and by providing direct instruction in the skills they will need to succeed as adults. They also need an overall compass or sense of direction so that they know what kind of human they should become. The culture of each society must fill these learning needs.

The culture of a society, or its collective psyche, is the collection of the learned behaviors and ideas of a group of people. It must shape the psyches of its individuals in such a way that a balance is struck between the needs of individuals and the needs of the group. Even within a single mind, conflicts can arise, but as other individuals get involved, the situation becomes even more complex. These conflicts are often resolved by discounting some of the needs. In capitalist societies, we have discounted the needs of the collective. In communist states, the needs of individuals are frequently subverted. In many patriarchal societies, the needs of women are ignored. In hierarchical societies, the needs of the elite rank highest. In many traditional religions, the needs of the old brain are scorned. In a hedonist's approach to life, the needs of the old brain are considered first. All of these extremist approaches are gross oversimplifications of a fundamentally complicated problem that may require millions of years to resolve. Any good solution will contain the potential for satisfaction of a broad range of the individual and collective needs.

Much of the conflict between the individual and the collective stems from the creation of mass societies of humans. Mass society has tremendous economic, military, and intellectual advantages. However, it is a new invention for us. It has only been within the last several thousand years that any sizable fraction of humanity lived in such an environment. We are only beginning to figure out how to elicit a social conscience that will work in this new environment. This new approach cannot depend on kinship or strict reciprocity because neither is applicable to total strangers. A system of law can cover gross behavioral offenses if criminals can be caught reliably. Furthermore, if the legal system is extended into the fine nuances of human behavior, it yields a sense of well being only for the lawyers. Belief in an afterlife in which the scales are balanced in the end only works for those who believe. An additional way to elicit a social conscience is to teach an appreciation of the beauty of others, and to see a kinship of spirit in them. This will hopefully lead to a reluctance to hurt or degrade other people in the pursuit of selfish ends. As with many things, the diversity of humanity calls for an answer that includes all of the above, and more.

The conflict between the needs of the individual and the group, or other individuals, is not new. Mass society is new. Sixty million years ago, our primitive ancestors were members of small bands living in trees, foraging for insects and vegetation. As with elephant families, behaviors that enhanced the survivability of the group enhanced the survivability of the genes of the individual. Therefore, some altruistic behaviors had a survival advantage and evolved naturally. In small societies there are many mechanisms based on kinship, reciprocity, and reputation for generating altruistic behaviors.

The dependence of the individual on the group became much more intense about seven million years ago, when our early ancestors first came out of the jungle and onto the savannas and sparse woodlands of eastern Africa. Then, since the trees of the jungle had provided safety from many of the predators, collective defense became imperative. Membership in the troop became more critical to survival.

The motivation for this dangerous move out of the safety of the jungle came from deep within the earth. The rise of the Tibetan plateau and the Rocky Mountains led to a drier climate. Across much of Africa, the jungles gradually faded away and were replaced by savannas. The Great Rift Valley that extends from the Nile River and Lake Victoria south to the Indian Ocean also began to form a natural barrier at this time that separated these two environments. A mountain range rose on the western edge of the rift. It squeezed the moisture out of the air onto its jungle-covered western slopes. To the east of the Rift Valley the climate was dry. Our closest relative, the chimpanzee, still lives in the equatorial jungles in the west in much the same way that our ancestors did. It lives mainly on fruits and other vegetation but will eat meat when it can get it. To us it appears to occupy the Garden of Eden. However, warfare and cannibalism have been seen there as well by Jane Goodall [1].

At the height of the glacial periods that have waxed and waned over the last several million years, the idyllic jungle environment of the chimpanzee almost completely disappeared. Its lack of adaptation to the new, drier environment brought the chimpanzee perilously close to extinction when its habitat shrunk from 40% of Africa down to only 3%. On the open savannas and sparse woodlands that constituted the majority of Africa, our ancestors had a hard life. It is unlikely that they engaged in any significant hunting at first. They were not fast enough and they had no tools appropriate to hunting. They probably were scavengers, competing with hyenas and vultures for the leavings of lions, sabertooth tigers, and cheetahs. Because of their ability to climb trees, they could also steal the kills of leopards. When these prehumans first came out of the rain forest, their only tools were unmodified sticks and stones to break bones. Chimpanzees have been observed to use these in defense, though their competence with them is very limited. They will throw them at a predator but the effect is more to startle and confuse the target animal as opposed to wounding it significantly. However, a farmer in east Africa was recently killed by a troop of rock-throwing chimpanzees.

At first, the chief use of stones was in butchering carrion. Early hominids did not have the sharp teeth or the bone-crushing jaws of the hyenas with whom they competed for carrion. Sharp-edged stones were used to cut the hide and to sever the ligaments of large animals. Two large stones, used as a hammer and anvil, enabled these hominids to crush bones in order to get at the marrow inside. The brains of an animal could be accessed with a single large stone. These are parts that only the hyena could also utilize. Therefore, the use of found stones allowed hominids to recycle the parts of lion kills that few other animals could exploit.

The ability to get at the fat-rich marrow and brains also provided a counter cyclic food source. In lean times, the bodies of animals that had starved to death were abundant. Our unhealthy taste for fat is probably a throwback to this era when our ancestors had a hard time getting enough of it. It was a lot of work to get at the marrow, so it had to taste good in order to motivate the activity. Recent studies show that about two hours of work with unmodified stones can yield one meal of marrow.

Between seven and three million years ago, the shape of our ancestor's body had evolved to be very similar to our own, even as its brain had remained small. Its body was the size of a modern pygmy (just under five feet high for an adult male). In relationship to body size, its brain was about one third the size of ours. It walked upright but like the chimpanzee. It had a small bottom, short big toes, long middle toes, and short thumbs. Many different variations of Australopithecus coexisted 2.5 million years ago. The Robustus line consisted of large, tool-using vegetarians. One species of these apes would dwarf the modern gorilla. This line ended mysteriously a million years ago.

The use of found, unmodified stones as tools went on for millions of years before hominids became tool makers. The first tools that were made consisted of stones that were flaked (shaped) only on one side. These tools had only one or two flakes (chips) removed from only one side of the original stone. This is known as the Oldowan culture. They were used to cut and to scrape. They appear in abundance about 2.5 million years ago. Soon after the start of this culture, Australopithecus evolved into the larger-brained Homo habilus.

The coincidence between tool making and the increase in brain size suggests that the mental demands of tool making and the use of made tools drove the evolution of larger brains. The line of omnivorous, tool-making scavengers that evolved into us spent a million years, flaking tools on only one side. These hominids radiated out of Africa into Asia. About 1.5 million years ago they evolved into a more advanced form called Homo erectus. This ancestor can be viewed as almost fully human. It stood about five feet tall and had a brain that was about two-thirds the size of ours. Roughly coincident with the appearance of Homo erectus is the development of stone tools that were flaked on two faces. The culture that produced these more complicated tools is known as the Acheulean culture. A very common tool in this new kit was the Acheulean hand axe. It is about six inches long and is flaked to a point. It first appeared about 1.5 million years ago and was used until relatively recent times. In addition to the production of more complex tools, Homoerectus also acquired the use of fire for warmth, cooking, hunting, and protection from predators about a million years ago. Though it is not clear when they could start a flame [2] they certainly knew how to keep it going. A cave in Asia has an ash pit that is forty feet thick with ashes.

A few hundred thousand years ago the evolution of Homo erectus resulted in two distinct large-brained humans. The Neanderthals had brains that were slightly larger than ours. They lived in Europe and parts of Asia. They were barrel-chested, bull-necked, and had curved thighbones. They were very strong by comparison to modern humans. They buried their dead, or just the heads, and often flowers were placed in the graves. These rites suggest the existence of a mythology. Some kind of language would have been needed to communicate such a mythology to the next generation. However, it is not clear whether or not they could generate the complex sounds that are characteristic of modern languages. The structure of their larynx appears to have been closer to that of the chimpanzees (who cannot speak) than to modern humans. The last of the Neanderthals died out in Europe about 40,000 years ago after being driven into isolated pockets by our ancestors.

The branch of the human family tree known as Cro-Magnons is the one that led to us. They were gracile by comparison to the Neanderthals. Genetic studies indicate that they originated in Africa about two hundred thousand years ago [3]. From there, they radiated into Europe and Asia by a hundred thousand years ago. Their tools were more sophisticated than those of the Neanderthals and were soon copied, in part, by them. Later, Cro-Magnons were definitely hunters. By 60,000 years ago, 40% of the big game species in Africa had disappeared. The entry of our ancestors into Europe was soon followed by the extinction of 50% of large animal species. After modern humans entered the New World about 12,000 years ago, 70% of large game animals went extinct. It is interesting that the only large game in the Americas that exist today migrated from Asia at about the same time as humans; they had already evolved the fear, speed, and stealth that allowed them to coexist with human hunters [4].

The moral of this story of destruction derives from the fact that our ancestors thrived by eating everything in sight. This is a dark side of human nature that will continue to cause destruction until we see fit to change it. It is against our economic interests and the value associated with the existence of diverse life forms to wipe out any. Our ancestors did not know what they were doing, but we have knowledge and the existence of alternatives. If we destroy a species that has suffered a billion years of life-and-death struggles to shape its being, then we are guilty of a severe moral offense against it and all of its ancestors. It is also an act of economic stupidity. The scientific value of the last remnants of a species that was shaped by tens of millions of years of struggle far exceeds the value of the few meals that their bodies might provide. The value of the diversity that it provides cannot be calculated.

The development of hunting tools and skills over the last several hundred thousand years has been paralleled by the development of arts, culture, language, and mythology. Bead necklaces became common eighty thousand years ago. The presence of four thousand beads in the grave of a young girl suggests economic/social stratification. Only a rich man could have afforded to bury the thousands of man-hours of labor that went into fabricating these beads. At present minimum wage, such a necklace would cost twenty thousand dollars to produce. This and other grave goods also suggest a mythology involving sacrifice and an after-life.

Drawings on the walls of caves in France from forty thousand years ago appear to have been part of a ritual of some kind. Many are so deep within the caves that visitors of the day would have had to use primitive lamps to find their way. Beginning about twenty-five thousand years ago, small statues of bulbous women, known as the "Venus" figurines, appeared. They were often worn on strings around the neck. Recent analysis indicates that on some of them, the sculptors carved representations of woven clothing. Over 200 have been found in Europe and Eurasia. The great prevalence of these statues, their form, and their context might suggest a widespread fertility cult. The existence of these artifacts, and their associated mythologies, implies complex cultures that required similarly complex languages to transmit culture from one generation to the next. Language may in fact be the characteristic that most distinguishes Cro-Magnons from their nearest relatives (Neanderthals and Homo erectus). Language's survival advantages, and its importance to cultural development, cannot be overstated. Speech enables people to expand their perceptions from the here-and-now to the everywhere and to all of time. It enables people to pool their knowledge and concepts. In the process, new knowledge and concepts must have been synthesized that would have been previously inaccessible to any individual. It did for the world of mind what sexuality did for the biological level of existence. Ultimately, language has united all of the human minds on this planet into a single collective psyche. This new level of reality is the most profound and complex evolved creation of the Big Bang. It provides the medium within which mental entities evolve and proliferate in processes that are analogous to the phenomena associated with replication and evolution of biological life. It is one of the most important forces in the shaping of modern humans and therefore cannot be neglected in this discourse on human spirit.

The use of complex language not only made it possible to transmit utilitarian knowledge from one human to another; it also made it possible to communicate a deeper understanding of individual psychological essence. Before this facility, people had to learn from each other by example and through rudimentary signals. Mothers, siblings, and nearest relatives were the primarily source of the psychic component of being that goes beyond DNA. With the advent of language, concepts could be transferred between any two. The influence of mentors, teachers, friends, and lovers became more important. These are examples of the horizontal propagation of psychic content within a generation. The vertical propagation from generation to generation also became more complex.

The invention of writing and mass media has made horizontal propagation even more important. Now many modern individuals may identify more strongly with their generation than they do with their family. The influences that shape individual psychological growth have diffused out across the entire collectivity of human minds. Our psychic being is imbedded in this collectivity of mind more deeply than ever before. The biological being of an individual animal arises from the gene pool of its species. This biological aspect of being returns to the gene pool through sexual reproduction. The psychic being, or spirit, of an individual stands in analogous relationship to the collectivity of human personality. There is a kind of immortality, or reincarnation, in this process of propagation of mental content from one individual to another. But as with the survival of individual genes within the gene pool, it is fragments of the individual psyche that go on in this way, and not the unique whole self.

The evolution of complex language out of the relatively simple vocalization and visual signals of lower animals is one of the most important stepping-stones on our journey to self-reflective consciousness. The depth of understanding and knowledge, necessary to either write or read this page, is beyond the grasp of any truly isolated individual. It is only through the evolution of our collective knowledge that we can approach these issues. The existence of this body of human knowledge, upon which we all depend, and the "evolutionary" processes that occur within it, depends on language. It is like the chemical foundation from which biology sprung into being. It is one of the most important forces in the shaping of the spirit of modern humans.

Just how and when complex speech arose may never be resolved. The soft parts of the human larynx do not fossilize readily. The bone structures that surround the larynx cannot produce unequivocal evidence of vocal ability. Reconstructions of the larynx of Neanderthals have been used to assert that their vocal abilities were closer to that of the speechless chimpanzees than to our own. Studies of a small bone that is part of the larynx have produced similar conclusions. Perhaps some day, a Neanderthal or a Homo erectus will turn up in a well-preserved state in a peat bog or a glacier. Until such evidence is available, the debate on whether or not complex language is a distinguishing characteristic of Cro-Magnon humans will continue.

In addition to evolutionary changes in the vocal apparatus in the larynx, changes in the brain are also required to form and recognize the complex sounds that are used in human language. Modern languages use over a hundred different sounds that are part of an even larger "hard wired" set that children have access to. Adults loose the ability to recognize and form the sounds that are not used in their native tongues [5]. Even two month-old infants respond to these sounds [6]. The parts of the temporal lobes of the brain used to form and understand complex sounds are even more invisible than the larynx to archaeologists.

Alternate lines of evidence can be derived from studies of language in modern humans. The hypothesis that all of the hundreds of different modern languages evolved (like DNA) from a single ancestral tongue can be used to estimate (with large errors) the age of this process. This is done by cross-comparing the commonly used words of different languages. This type of study indicates that modern human languages have been evolving for at least one hundred thousand years from a root tongue. Studies of the variability of DNA in modern humans indicate that we evolved from a small group of ancestors that lived two hundred thousand years ago. These numbers are consistent with the fossil record that indicates a branching of Cro-Magnon from Homo erectus at about that time in the past.

The fact that biological mutations were necessary for the use of complex speech has been demonstrated by numerous attempts to teach apes to speak. All of these efforts have failed because of deficiencies in the structure of their larynx. They simply cannot make the range of sounds necessary for modern languages. The lack of talent in this area is not due to mental deficiencies. After many studies, and a lot of controversy, it is becoming clear that a wide range of apes (chimpanzees, gorillas, and orangutans) are proficient and creative in their use of sign language and other visually-based languages. They carry on charming conversations with people and other apes. The content and style of their communications bears an uncanny resemblance to that of a three-year-old child. Whether or not these animals have mastered the nuances of syntax (a common focus of criticism) does not diminish the fact that when they are provided with appropriate tools, they create complex communications that go well beyond what happens in the wild. A number of these studies are now looking into whether or not signing mother apes will teach their babies to sign. Given a few million years, these close relatives might evolve into a second line of self-reflective creatures.

Looking back on the forgoing discussion, we see that in the last seven million years, a very close relative of the chimpanzee evolved into an erect biped that used found tools, then into a large-brained tool maker, and then into a larger-brained hunter and user of language. By the end of the last ice age, fourteen thousand years ago, fully modern humans were the only ones left. Whether Neanderthal and Homo erectus succumbed to the harsh climate, economic competition, or military conquest, is not known. Contrary to genetic studies, many paleontologists maintain that interbreeding assimilated these diverse forms into a single human species.

In any case, it is clear that the humans that were around when the climate improved were very good at hunting and surviving. In fact, they were too good at it. As pointed out above, everywhere they went big game disappeared. Slow animals that had survived the predation of non-human carnivores were an easy target for tool-wielding humans. They could bring down anything with fluted spear points. They went extinct before their genes knew what hit. Within fifty thousand years people had wiped out most of the economic base that had made them so successful.

As with many disasters in our history, the extinction of our favorite big game animals led to a new spurt of creativity. As hunting became more difficult, people adapted in various ways. The Laps, who live in Europe near the Arctic Circle, developed a possessive attitude to the herds of wild reindeer that they previously hunted. Though these deer are not domesticated to anywhere near the extent of the modern cow, they are herded, milked, and used to pull sleds. In the middle east, people also began to domesticate pigs, cows, sheep, and goats. They gathered grain from the wild grasses that were abundant in the region. By ten thousand years ago people in the Middle East had learned to plant wheat, barley, oats, and flax (for linen cloth). At around this time, rice was being planted in some regions of China. The agricultural revolution had begun. Civilization, commerce, hierarchy, writing, the arts, technology, and science evolved in many places where agriculture took route.

The great economic success of agriculturally based societies numbered the days of the hunter-gatherers that went before. They could not compete economically or militarily. The high population density that could be supported by agriculture depleted game well below the density that hunters required. In Medieval Europe, the nobles reserved the remaining game for their sport. The high population density also represented an overwhelming military advantage. Everywhere the farmers went they took their language and culture. The hunter-gatherers melted away before them. Studies of the spread of languages and agriculture indicate that, in general, the two went together. This appears to be true for the migration of agriculture out of the Middle East and into India and Europe. It is also the case for the island-hopping spread of agriculture across the Pacific Ocean starting in South East Asia. The pattern of interrelationship of the languages in these regions matches the spread of agriculture.

The way in which people lived and worked together changed radically with the advent of agriculture. What a happy and successful human being had to be changed forever. The free spirit of the nomadic hunter-gatherer had to evolve into the sedentary and constrained member of farming societies. Even by the relatively rapid evolutionary time scale of the learned aspect of human psyche, this ten-thousand-year change occurred overnight. On a genetic time scale it was in the blink of an eye. Both aspects of human spirit will be on an evolutionary fast track for some time into the future in an effort to adapt to the advent of worldwide mass society.

In the new medium of small agricultural societies, mass societies arose through military conquest. Small, nomadic, hunter-gatherer tribes do not make good subjects for a king. Among other things, you never know where to find them when it is time to collect taxes. In general, their foods are eaten on the spot and cannot be preserved on the long journey back to the center of power [7]. In short, being a producer instead of a parasite was more adaptive before the advent of agriculture. In the last ten thousand years the tables have been turned by the horn of plenty that agriculture has produced. The learned aspect of human spirit is rapidly evolving to this new reality. Almost every one of us would find the hunter-gatherer lifestyle unbearably demanding even if we knew all the things that primitive people do, if our physical conditioning was up to it, and if an abundant natural environment could be found on this planet.

In the last half of this century, almost all of the remaining Paleolithic societies have succumbed to the seductions and intrusions of modern life. The Eskimos of Alaska have given up living on the ice pack in winter, as they did for tens of thousands of years. North American Indians have been forced onto reservations where running casinos is the best business opportunity available. The Yanomamo of the Brazilian rain forest are being dragged into the twentieth century whether they like it or not. Though their males previously suffered a 30% mortality rate from intertribal warfare [8], this may have been preferable to their children dying of the diseases that came with the encroachments of gold miners and the water pollution that they created. In the remote highlands of New Guinea, people whose tool kit was based on wood, fiber, and stone, now prize the one steel axe that the village possesses.

The ability of the easy life to corrupt the wild spirit is not limited to humans. The difficulties associated with returning orphaned chimpanzees to the jungle is leading to despair among conservationists in Africa. Over 200 orphaned chimps have accumulated in shelters in the last twenty years. None have been successfully reintroduced into the wild despite numerous sincere and intelligent attempts [9]. Like it or not, almost no one is going back to the garden. Attempts to form agricultural communes (not even hunter-gatherer ones) in the

1960's and a hundred years before have almost all ended in failure. However, for many of us, the health of our spirit still demands that we be able to "return to the garden", at least on occasion. We can vacation in semi-wild areas. Or we can watch a nature program; just knowing that such places still exist can be comforting.

Though it is easy to be nostalgic about the loss of life in the wild, and the loss of associated aspects of human spirit, most of us cry crocodile tears over the issue. The material and psychic compensations of modern life out-weighs the loss for most of us. Therefore, we rush headlong into an ever more rapidly-evolving new world. Agriculturally based civilization provides a new medium in which innumerable collective mental entities can evolve as never before. The structure of an agriculturally based village evolved out of the structure of a hunter-gatherer tribe. Soon political structures evolved into hereditary kingship.

The rise of vast civilizations led to many new technical and psychic inventions that had never existed before. Political collectivity produced mass societies in which the old moral systems, based on kinship, reciprocity, and reputation (face), were less effective at preventing sociopathic behavior. Systems of law were invented to allow society to function in an orderly fashion. Religions evolved to provide mythological foundations for laws in an effort to shore up societies. Within stable societies that were often protected from invasion by natural barriers, commerce, industry, agriculture, and technology evolved toward ever higher levels of development. As society became more complex, the need for records of commercial transactions became more acute. This need was the mother of the invention of writing. The development of writing is one of the great stepping-stones to our modern technical civilization and to the philosophic and scientific self-reflectivity that goes with it.

As with language itself, the importance of writing cannot be over-stated. For the first time, complex histories could be preserved with accuracy or inaccuracy. The awareness of the self, on a societal level, got a tremendous boost through accurately recorded history. New religions based on written text developed a level of conformity and dogma that was previously unheard of. Some of these new religions became the bone of contention in many horrible wars.

Writing enabled scientific and mathematical developments to be preserved across the times of political crisis (e.g. the Dark Ages) when there were no practitioners of these disciplines. It has also been like a bridge across times when there was no one with enough genius to make the next breakthrough. Without written records, the experimental foundations of modern science would be impossible. There would be endless disputes (that the scientific method is supposed to resolve) about what really happened. The development of complex theories to explain the data would be impossible without the written language of mathematics to describe them.

Written language enables individual authors to project their ideas, and their sense of what human spirit is, or should be, across the entire globe and to all future generations. A large part of the being of Shakespeare (and other great authors) will be forever with us to point out follies and the genuine joys of being human. The proliferation of available literature from a huge range of authors was a great gain for the collective psyche of humanity.

With written orders, rulers could control empires that required months to traverse. Empires could therefore expand to cover significant fractions of the globe. The first truly mass societies came into existence. Within these societies, new ways of doing things could evolve and propagate. Systems of written laws and procedures evolved. Bureaucrats were invented.

With written records individuals could run vast commercial empires. It was this use of writing that first got things going in Mesopotamia many thousands of years ago. When a herd of animals was being driven from one town to another, it was necessary to have a tamper-proof record of the number of each kind of animal to prevent drivers from skimming off animals. Hollow clay balls that contained tiny figurines of each animal (sheep, goat, pig, etc.), were sent with the herd. On the receiving end, the ball was broken open and the tally of figurines was matched to the herd. A later development was to scratch pictures and symbols of the animals on the outside of the ball, as well. This kept things honest at every transfer point without having to break open the ball and make a new one. It was a small step from this to scratches on clay tablets, what we now know as Sumerian writing. The generalization of this to writing on the walls of tombs, sheepskins, and papyrus paper was an easy step.

The use of writing quickly expanded from mundane commercial applications to political, historic, religious, scientific, technical, and literary spheres. It created new opportunities for evolution everywhere. The world of collectively held knowledge and ideas made quantum leaps in development that are still occurring today. The shape of the psyche of each individual became ever more influenced and imbedded in this expanding collectivity. The nature of human spirit began a new round of evolution based on written inheritance (e.g. the Bible, the Constitution, etc.) as opposed to oral traditions. Mass media would ultimately become a dominant force in the way people thought and behaved.

The technological revolution has its roots in tools used by our ancestors millions of years ago. From two-and-a-half-million years ago until the dawn of the agricultural revolution, the raw materials of technology were stone, wood, bone, skin, and fiber. The pooling of knowledge, the leisure, the new needs, and the stable home of the agricultural peoples generated a new spurt of technological innovation. Clay was fashioned and fired into ceramic vessels for storing grain and cooking. Clay bricks (both fired and unfired) were used for construction. Flax was spun and woven into linen cloth. The wheel was invented for carts to haul goods. Large boats were fashioned for transportation and commerce on rivers. As boats evolved towards greater sea-worthiness they moved out onto the coastal areas, the open Mediterranean and ultimately the major oceans. All of these developments were of great importance but they pale by comparison to the development of the technology of metals.

The use of metals for tools and jewelry ushered in innumerable innovations. The use of metals for jewelry preceded the agricultural revolution in many places. However, the development of wide applications for metals occurred in agricultural societies. The ore deposits, smelting ovens, and fabrication facilities required geographic stability. A nomad would find it prohibitively difficult to truck these things around while looking for the next meal. Many metals, including gold, silver, and copper, can be found in ores as native metals and do not require smelting before they are used. They can be found as relatively pure metals that are not chemically combined with oxygen or sulfur. These native metals were the basis for early metal-working industries.

In addition to existing as native metals, gold, silver, and copper are malleable enough to be worked at room temperature. Using a hammer and anvil pieces of these metals can be pounded into many shapes without breaking the object. This same pounding technique can also be used to cold weld small objects together to make large ones. This technique of working soft metals was developed over thousands of years and was applied to jewelry, gold leaf decoration, vessels, needles, and other small specialized high value tools. Because they were so labor intensive, only the rich could finance these industries. Commoners still relied on the traditional raw materials for their tools.

Metalworking might have remained a niche technology if it were not for the development of the smelting of sulfide (and oxide) ores into copper, and then bronze. The initial discovery of bits of copper in the ashes of a campfire eventually lead to the using of the same kilns for smelting that were used to fire clay. Contamination of some ores with tin or arsenic led to the discovery of bronze and to the real start of the age of metals. Bronze is an alloy of copper with these other metals and is much more useful than copper first because it can be readily melted and cast into complex shapes with primitive kilns, and second, because it is much harder and stronger. Bronze axes, knives, swords and armor soon became the preferred tools of warfare.

The Bronze Age gave a tremendous impetus to hierarchical society. Prior to this, everyone had access to the tools of warfare. However, these wood and stone tools were no match for bronze swords and armor. Only rich men could play. Empire and slavery became common place. The rich got richer and the poor got poorer. Human spirit took another turn in the direction of elitism. Four thousand years of cultural and technical evolution would be required to reverse the trend. Mechanization has now removed much of the economic incentive for slavery.

The Bronze Age lasted a couple thousand years until the Iron Age. Though iron could not be melted and cast the way bronze could, it was twice as strong and less prone to fracture. It therefore became the preferred metal for armaments. Because of its higher melting point, it was much more difficult to win from the ore and to work into tools. It was therefore more expensive. An average suit of armor would require a man-year of effort from a skilled blacksmith. Only the sons of the rich could be knights.

Perhaps this is painting too dark a picture of the age of metals. The Aztecs managed to support a very elitist and brutal society with stone, wood, and fiber. Their battleaxes had wood shafts topped by a razor-sharp head made from obsidian, a volcanic glass. Their armor was so light weight that it was preferred by some of the conquistadors.

The bright side of metals comes from the fact that the same technology that can beat a sword into shape can also beat a plowshare into shape. Continued evolution of these technologies ultimately led to the creation of modern machinery for performing innumerable tasks better and cheaper. With each succeeding generation, the comforts of the rich have gradually diffused down to everyone. Television sets can now be found in the most humble hovels in the third world. The laborsaving machines of the modern middle class American family will soon be found everywhere in the world.

The lifestyle that was available to a king in Syria, four thousand years ago, is now available to middle class Americans. Clay tablets from that era record the provisions of food, beer, and wine for a big bash with neighboring nobility. The list could easily be mistaken for a modern shopping list for a lawn party. You don't have to be a slave owner to have a good time anymore. If you need a ditch dug, it is not necessary to have a lap-dog army that watches over a bunch of slaves. A backhoe is now the efficient way to do it. Similarly, a washing machine does a better job than a slave slapping the clothes against a rock.

This emphasis on the material benefits of technology is not an effort to boost materialism as opposed to the psychic aspect of

existence. It is merely a frank acknowledgment of the proven his-
torical fact that people will exploit each other if that is what is necessary in order to satisfy their material needs. Maslow's "hierarchy
of needs" formalizes the concept that unless people are adequately
fed and housed, they are incapable of thinking of higher needs. The
fact that technology enables these lower needs to be fulfilled (without excessive exploitation of others) can have a profoundly positive
impact on the evolution of the human spirit. It is unfortunate that
modern commercialism leads to the creation of ever more material
needs and obsessions. It is a choice, on the part of people in the
commercial arena, to create these artificial material needs as fast
as new technologies can satisfy them. The rat race is voluntary. It
is not dictated by the existence of technology. Technology simply
makes this shallow approach to life more possible without resorting
to slavery.

In addition to satisfying our material needs, modern technology
has created the ability to transport individuals and information all
over the globe at a cost that is almost negligible to middle class
Americans. We now have Marshall McLuhan's global village or
Pierre de Chardin's Noosphere. Everyone can be in some kind of
physical or psychic contact with everyone else. The rate of technological growth is accelerating because developments in any one of
the high-technology societies quickly diffuse to them all. The free
exchange of ideas is also leading to astounding rates of cultural evolution as well. The brutality that has characterized many societies
in the recent past is becoming very much out of fashion. This is due
to the focus of world attention that can occur through mass media.
The repressive regimes that were spawned by Stalinism have melted
away in the medium of the free exchange of information. South
Africa is now run by the majority.

Through visual media, the character of a remarkable individual like Mohandas Gandhi can be viewed by everyone. His ideals can be assimilated and become part of the spirit of all of humanity. He said, "We must all be the change we wish to see in the world." He was this, and he made it work to a great extent. The interconnectedness that modern technology creates allowed visible change in a half century. The impact of Jesus' life took thousands of years to settle in. Even that depended heavily on written media in the form of the Bible. In the future, the evolution of human spirit will occur in the media as well as in individual human interactions (the way of the past). This is both good and bad. The tremendous accessibility that is provided by the media is counterbalanced by the fact that comic book images are what is most easily transmitted.

Having boosted some of the Yang (the bright side) of the cultural, political, spiritual, and technological developments of the last ten thousand years, it is time to visit the Yin. Though the high rate of change has many positive effects, it is also a severe problem as well. In any evolutionary process it is important that things be tried out and refined before the next innovation is thrown in. As things get more complex this is even more the case. Cellular life that used DNA as its means of storing the genetic wisdom of the ages won out over RNA-based life forms. This is because DNA mutates less rapidly. Right now, the rate of change in the way people structure their minds and lives is so fast that pathologies are evident everywhere. The crime rate, the incidence of single-parent homes, and the suicide rate are a few indicators. We are not necessarily becoming happier people. We are losing a sense of what produces genuine value and happiness in life.

Along with the deterioration of the psychic ecology of humanity is a deterioration of the biological ecology of the planet. Agricultural and medical technology has allowed the population to expand past sustainable limits in many parts of the world. At the same time, the insatiable material pursuits in the developed world are producing an unconscionable level of resource expenditure and pollution. Modern transportation has allowed AIDS to spread across the globe before anyone even knew it existed. This is not a new story, however. Shipping allowed the Black Death to spread across Europe from the Black Sea within a few years. Also, a hundred million native Americans died of disease within a century of Columbus' voyage. The Yanomamo of the Brazilian rain forest are being decimated by modern diseases today. The rain forest in South America and Africa are visible at night from space due to fires set for slash and burn agriculture.

This slash-and-burn agricultural technique converted the tropical paradise of Easter Island into a waste land. A thousand years ago, a statue-carving, ancestor-worshiping culture flourished there. Hundreds of stone busts of ancestors dot the island. They stand twenty feet high and weigh more than twenty tons. Just how they were carved and dragged to the sights where they stand is still being studied. After the last tree was felled and the last field farmed out, their economic base collapsed. The people could no longer fish and their fields were depleted. Easter Island turned into a wasteland inhabited by cannibals. A population of ten thousand was reduced to a few hundred by the time Captain Cook arrived. There are lessons in all of this for the collective thought process of humanity in the twenty-first century. If we meet the challenge of controlling population and resource expenditure, then we can avoid the living hell that Easter Island became. The form that the human spirit takes two hundred years from now depends on how we deal with these challenges. If our folly turns our environment mean, then our spirit will evolve to match it.

Our Improbable Universe

Modern technology has hung yet another Sword of Damocles over the collective fate of humanity. This sword is the existence of nuclear weaponry. If human spirit on this planet does not evolve into a form that can coexist with nuclear weaponry, then it will cease to exist. There is no way that we can repeatedly sally-up to the precipice of all out nuclear war (as we did during the Cuban missile crisis) and come out alive. Each time is like playing Russian Roulette. Over the centuries, the odds will catch up with us. Our only hope is for the nature of human spirit to evolve to a form that can avoid their use.

Over the last ten million years we have seen a fun-loving ancestral ape that lived in an African jungle evolve into fun-loving members of a world wide collective mind, economy, and ecosystem. The psychic nature of that ancestral beast evolved in coordination with its DNA to make us what we are today. We now have the degree of self-awareness that we need to chose the future form of human spirit. If we make the right choices (the adaptive ones), then we may succeed in dodging the many bullets flying towards us. If we fail to do this, then perhaps, in the wide open planet that results from our failure, some other animal will do so. Before a senate committee, Admiral Richover (the father of the nuclear navy), in a moment of dark humor, suggested that it might be the humble rabbit that ultimately succeeds. There is still another billion years left to this planet before the gradual heating up of the sun causes a run away greenhouse effect that converts the ocean into steam and the planet into a facsimile of Venus. Maybe the highly evolved descendants of the rat will inherit the Earth and save it from this fate. Personally, I'd rather it be our descendents that become the stewards of this beautiful and diverse planet.

Chapter 11
People Are Creators

We have seen how ascending levels of complexity have evolved the raw energy of the Big Bang into the biology and the spirit of humanity. It has also evolved into all that we have created. We are now the primary creators in this corner of the universe. We even stand at a threshold where humans will soon create biological life in a test tube based on self-replicating strands of RNA [1]. This development was predicted by Mary Shelley in the story of the Frankenstein monster one hundred and fifty years ago. Whether or not these efforts succeed in creating biological replicas of life, people have already created many life-like entities. These creations are not made out of biological chemicals but out of human thought. In many cases, the things that have come out of the collection of all human minds (our collective psyche both conscientious and unconscious) have assumed a life-like character of their own. In this sense we are creators of new forms of life.

Ideas, myths, inventions, cliches, economic systems, political systems, corporations, languages, fashions, religions, scientific theories, jokes, games, states, computer programs, computer virus, etc., are entities that evolve and replicate (propagate) in the medium of all human minds [2]. They shape the human mind and are shaped by it. Individual humans are partly made out of these things and these psychic entities arise from the collection of individuals. Ideally, they interact with people in a symbiotic (mutually beneficial) relationship.

Unfortunately, the relationship is often parasitic (e.g. a fascist state oppressing its people). Their totality forms a kind of ecology. Many have suggested that the human collective psyche will ultimately couple to an interstellar network of thinking beings and thus generate yet another level of psychic complexity in which life will evolve for as long as the universe exists [3].

The forms that these entities evolve into are determined by a process that is similar to biological natural selection. The versions of a particular entity that are most adaptive because they best fulfill some kind of human need are the ones that tend to propagate most readily into the next generation. As time goes on, a short campfire story about an abduction evolved into an epic myth like the Odyssey. A log used as a roller to move a heavy stone evolved into a wheel and then into a race car. Barter evolved into a modern capitalist system, which in turn forms a new medium in which other entities, like corporations, can evolve. And alchemy evolved into chemistry. The net result is greater fulfillment of intellectual, emotional, or economic needs and the emergence of institutions of unprecedented power to fill these human needs.

This analogy with biological evolution is a powerful way to understand how these psychic entities came into being. However, as with all analogies there are severe limitations. Evolution is only one way that human progress can be made. At times there is no way to get from here to there in a stepwise fashion. Often imperatives do not allow enough time to try out each step before the next is taken. At these times, revolutions become the mechanism for change. In revolutions, humans invent within their own minds a concept of how things should be and then execute that conception. A major change then occurs almost overnight, and it takes historians fifty or so years to figure out what happened.

The principle of creation through evolution can still be applied in the context of revolution, however. The mental process by which a revolution is conceived is also evolutionary. In the internal process of invention, many ideas are conceived, synthesized, and reality-tested before the final product is tried out. The process can be internal to an individual or the collective effort of a number of communicating thinkers. However, no one is smart enough to think of all of the angles. Therefore, it is inevitable that the product of a revolutionary concept will be full of bugs initially. These shortcomings usually require a good deal of evolution to correct.

Even in the realm of science and technology, it is hard to get it right the first time. All of science will remain an approximation to physical reality until the ultimate Theory of Everything (TOE) is found. The Nobel laureate, Sheldon Glashow, describes his approach to the ultimate theory as "bottoms up". He believes that the way to find the next level of understanding is to build it painstakingly on the previous. This is the traditional way—it is evolutionary. This approach can be contrasted to the theories that have a top-down approach, like Super String theory, an educated guess about the foundations of reality. The right guess, when mathematically unfolded, will be consistent with all that is already known and will fill in all that is not. The future will be the judge of who had the better approach.

Even though Einstein's theory of relativity appeared to jump out of nowhere, he in fact acknowledged his debt to Maxwell's equations of electromagnetism, written forty years earlier. As a young man Einstein became obsessed with his own insight that the constancy of the speed of light was implied by Maxwell's equations. He built on this to create what most people describe as a revolutionary theory. It created a revolution in thought but it evolved out of the work of

Our Improbable Universe

Maxwell, which in turn evolved out of the work of Faraday, Gauss, Ampere and many others.

In high technology it is well known that profits are hard to realize in the first generation of a new invention. Even after exhaustive testing, it is almost expected that bugs will have to be worked out of the product in the field and out of the manufacturing process that produces it. The first generation of commercial jet planes split open disastrously. The Chernobyl nuclear reactor was an old design that is inherently unstable. Though inherently safe designs exist (e.g. the high-temperature, gas-cooled reactor) they are not in commercial service because of the penny-wise-pound-foolish mindset of the industry. This is very unfortunate for the environment in that safe reactors will probably be used in the eleventh hour of planet Earth as an alternative to totally frying it. Hopefully James Lovelock's fear of too little too late will not come to pass.

As great as the need is for an evolutionary approach in well-defined arenas such as politics, economics, technology and the sciences, it is even more needed in the way that individual people structure their psyche. Here, it is much less obvious what the right approaches are. Despite this, every modern generation wants to throw out the approach of the previous one. Within half a century the extended family has become the nuclear family, which then changed into the divorced family or the single-parent family. In many cases, this progression arrived at the no-parent family, due to careers, drugs, or video games. The passing of the torch to the next generation is occurring through the media instead of through loving interpersonal relationships. Commitment and love for anyone else is out of fashion. Because the price in human happiness is hard to measure, the loss won't make headlines. These stresses on the ecology of human spirit are just a few of the many dislocations that the rapid pace of cultural revolution is precipitating.

The invention of written language was motivated by the need for financial records (see Chapter 10). Therefore, a great deal of economic evolution was driven by this invention. Written contracts (for all to see), corporations, international markets, advertising, and accounting all developed to unprecedented levels. Multinational corporations now dominate a complex world economy. The techniques for manufacturing and engineering high-technology products are rapidly propagating to every corner of the globe. Soon, all people will have access to the benefits of a modern economy. This will increase the number of people with a high standard of living by tenfold. But per capita pollution has kept pace. If we don't learn to generate the economic benefits without polluting the globe, then the down side of this expansion of the econosphere will be a seriously damaged biosphere.

The need for information processing for complex businesses spawned the computer revolution. In the half century that computers have been around, the processing and storage abilities of these machines has expanded more than a million fold. The Internet has now connected these machines and enhanced the connectivity of human minds. We are now joined in a way that individuals can freely communicate with each other all over the planet. A side effect is that a cyber environment now exists in which cyber "life forms" (i.e. the computer virus) can reproduce, propagate, and evolve.

In the field of chemistry, computers can be used to simulate the complicated chemical reactions that occur when chemicals are mixed together. In the distant future the detailed chemistry of the 500 enzymes that the E-Coli bacteria needs to live will be understood well enough to simulate its life cycle all the way through cell division. Once the go button is pushed, the simulated bacteria in the memory banks of the computer will proceed to grow, divide, and mutate. The computer programmer will be able to claim to have created a new life form that is behaviorally similar to the real thing. The difference is that the medium for one is a dish of biological nutrients while the other lives in a computer on electrical energy that comes from the Hoover Dam. It's even imaginable that the evolution of higher organisms would proceed from such a simulation if a suitably hostile and competitive environment were also simulated.

Less sophisticated forms of artificial life have already been played with by scientists in their attempts to gain insight into real life. Ecologists [4] have studied symbiotic (cooperative) and parasitic relationships with simple computer entities that can evolve into simpler forms that reproduce more efficiently.

Artificial life simulations have also provided insight into the question of how altruism could spontaneously evolve from dog-eat-dog natural selection. A game called The Prisoner's Dilemma [5] has been played through many life cycles [6]. These simulations yield a favored strategy which can be seen in the behavior of stickleback fish when they team up to avoid predators [7]. In the simulations, entities play the game with each other millions of times. Depending on the number of points they accumulate, they reproduce (are copied to new memory locations) or die (are erased from memory). Each entity is given a strategy, or genetic code, for playing the game. When it reproduces, its offspring inherit that strategy. Various mixtures of entities with cooperative or exploitive lifestyles can be tried out in this way to see what this form of unnatural selection ultimately leads to after many simulated generations .

In another type of computer life "game," the entities have a choice of whether or not to play with each other. If the computer entities remember the outcome of past encounters, then the wolves (exploiters) get frozen out very quickly, because no one wants to play with them. If the population is large, it takes longer for the wolves to exploit all of the potential victims before their history catches up with them. It would be interesting to see what would happen if the outcomes of each encounter were a matter of public record to all players. Undoubtedly the principles behind the traditional concepts of reputation and honor would become manifest. It would be harder than ever for an exploiter to do well.

Many other forms of lifelike computer entities have been experimented with. They are interesting because they shed light on the life processes that humanity is still struggling to understand. There is also a certain fascination with being the creator of something like life and watching the drama unfold. The concept of the computer virus originated as part of cold war skullduggery. Now the motivation is primarily mischief. Whatever the reason for creating these little Frankenstein monsters, like the Love Bug, there is much debate as to whether or not they are a life form. Presently there are more than two thousand different viruses infecting the world's computers. In an effort to rid computers of these renegade pieces of code, workers at IBM are looking to the immune responses of animals for guidance in dealing with these pests.

One of the major differences between computer viruses and biological viruses is the accuracy with which their code is transcribed during reproduction. The hardware of computers is designed such that the probability of a copying error is minuscule. It is the process of mutation combined with an environment in which natural selection occurs that allows biological entities to become more than what they once were. This is a key feature of true life.

It is possible for a computer virus to have the ability to mutate randomly, as well, if the programmer so designs it. This kind of virus would be very lifelike and possibly capable of developing in a helpful direction in an effort not to be selected against by anti-virus system programs. Many life-like computer entities have been designed with the ability to mutate. When the environment is designed to select for a feature, it spontaneously arises. Very complex computing tasks will be solved in this way by lazy programmers in the future. They will simply allow a program to evolve in a selective environment until it does all of the right things.

In some sense, what are known as neural nets are taught their behavior in an evolutionary fashion. This type of programming was partly inspired by insights into the way that nerve cells work together in the nervous system of animals. The brains of animals recognize particular collections of perceptions and generate appropriate responses. At this time, they can simulate nervous systems that have somewhat more neurons than a typical worm. But certainly, this will increase greatly, and the network's capabilities will expand. Right now this type of programming is being used for speech, voice, and handwriting recognition, tasks for which neurons are well suited, yet fiendishly difficult to program.

The combination of neural network programming and mutation, in an environment of non-natural selection, may some day produce a conscious computer that can be a servant or a master of humanity. Numerous science fiction stories have been written on this theme, most notably Arthur Clark's 2001: *A Space Odyssey*. That so much fiction has been devoted to the negative potential of such lifelike entities may ensure that the enslavement of humanity by a machine will not happen. If it ever does, then a mutating computer virus may become our ally in dealing with the renegade computer.

Another way in which people could possibly become the creators of pseudo-life-forms is to create mini-universes within our own universe. This type of activity has been speculated upon by Allen Guth, one of the originators of the inflationary model of the Big Bang. He has proposed that if the energy in a few ounces of matter could be focussed to a tiny point it might trigger an inflationary "Little Bang" universe inside of the black hole it produced. However, because it would be inside a black hole, we would not be able to tell what was happening. Life forms could even evolve inside of it that, at times, felt abandoned by their creator.

After fourteen billion years of the evolution of matter and half a billion years of the evolution of mind, the universe has produced a remarkable group of beings whose individual capacities and whose interconnectedness gives them the power to create in unprecedented ways. Not only do we create static objects like buildings and books, we also create things that evolve into more than they were initially, like the styles of architecture and literary ideas. This chapter has enumerated and detailed a number of these lifelike things that evolve. These entities live in the collectivity of interconnected human minds, the chemical media in the test tubes of our laboratories, the memory banks of our computers, and in individual minds.

The previous chapter detailed the evolution of human spirit to the point where it could create all of these static and dynamic entities and consciously participate in creating itself. If this universe was created, then we have indeed evolved into an image of the creator even though our achievements have been puny by comparison. If the universe was not created, then we are its god-like conscious creativity. In either case, what we create is our responsibility and the act of creating is one of the great joys of being.

Chapter 12
Who's Out There?

"When there is plenty of matter in readiness, when space is available and no cause or circumstance impedes, then surely things must be wrought and effected. You have a store of atoms that could not be reckoned in full by the whole succession of living creatures. You have the same natural force to congregate them in any place precisely as they have congregated here. You are bound therefore to acknowledge that in other regions there are other Earths and various tribes of men and breeds of beasts." Lucretius, first century BC [1].

This millennium marks the fourth centennial of the "atrocious death" [2] of a Dominican friar named Giordano Bruno. He was an early proponent of the Copernican heliocentric theory of the solar system. He also proposed that Earth might be one of an infinite number of worlds inhabited by other beings. For this and other beliefs, the Inquisition saw fit to condemn Bruno to be burned at the stake. Interestingly, though, he has been belatedly mourned by the institutional descendents of his persecutors. Cardinal Angelo Sodano has issued an apology of sorts, as many of his ideas are currently accepted. These days, the big question is not "Are we alone?" but "Who's out there?" What is the mental character of our nearest neighbors? Are they as intolerant of diversity as people commonly were up until the very recent past? Will they kill you if they don't like what you say?

Biological life forms on Earth take many forms, but they all share a common chemistry, based on nucleic acids, proteins, sugars, and fats. From the simplest virus to the most complex mammals, the genetic code has essentially the same structure. The difference is in the size and complexity of the genetic message passed from generation to generation. In four billion years, this message has evolved from a length of a few hundred parts to several billion. The functions specified by the "coded" message have evolved from simple cellular reproduction to all of the talents manifest in humans and the remarkable plants and animals with whom we share this planet. The commonality of the genetic code suggests that all of the biological life on this planet descended from a single primordial form. Under mysterious conditions these elements spontaneously assembled themselves into self-replicating strings of nucleic acids.

Whether or not human investigators succeed at creating artificial chemical life, the universe has likely done so many times over in its fourteen billion year history. As Lucretius correctly surmised, all of the chemicals that we are based upon exist in abundance in naturally occurring environments throughout our galaxy of a billion stars and throughout the universe of trillions of similar galaxies. The early environment on the Earth wasted little time in creating life here. The oldest fossils of primitive cells are almost as old as the oldest rocks. Therefore any environment like the infant Earth must have a high probability of developing from a chemical soup into some form of life. Recent observations by the Hubble space telescope of young stars in the Orion Nebula prove that planetary systems are more common than we thought. More than a billion trillion such systems exist in this universe. Therefore it must be teaming with life.

The kinds of life that exist out there in the far reaches of the universe may or may not be based on the chemistry of nucleic acids or even that of carbon. If alien life is found that is based on nucleic acids and pro-teins, the question will arise as to whether it evolved independently, or if it shares a common ancestry with us. Interstellar transportation of bacteria on fragments of rock that are blasted into space by meteors is very feasible. Bacterial spores that had been locked in the stomach of an amber-encased bee for 130 million years were recently revived; spores can lie dormant for more than a hundred million years. The existence of this form of interplanetary transportation of rocks has already been proven by the discovery that many meteors that have fallen to Earth have the same chemical signature as soils that have been analyzed on Mars and the Moon. Our ancestral microbes may have arisen on Mars and hitched a ride here on a rock that was blasted from its surface by a large meteor. It is even possible that an alien space traveler inadvertently or deliberately contaminated Earth with microbes four billion years ago. If there is a common origin for life in this region of our galaxy, then it will be possible to discern it from the genetic code of alien creatures—if contact is ever established.

If life on nearby stars does not have a common origin, then it is still likely that it will be based on the same atoms as our biology: carbon, oxygen, hydrogen, nitrogen etc. As pointed out in Chapter 5, these elements can combine to form an incredible number of differ-ent molecules. Not only do these elements have the inherent poten-tial to produce life but they also exist in great abundance in the clouds of gas and dust from which stellar systems condense. The amino acids and nucleic acids that we need spontaneously form in naturally occurring environments and are even found in meteors. Therefore, the spontaneous evolution of carbon-based life is to be expected under the right conditions.

The chemistry of the ninety-two elements in the periodic table is so rich that it is impossible to exclude the idea that other chemical systems might spontaneously evolve life in one of the billion trillion environments that this universe provides. Similarly, it is not possible to exclude life from other levels of material reality. It is possible that the chemistry of nuclear matter on the surface of a neutron star might produce creatures that live at a speed that is a million times faster and a density that is a trillion times greater.

In an even wilder flight of imagination, one could conceive of a life form that was based on the quark chemistry that existed within the first millisecond of the Big Bang. Such a life form could only exist in the super-dense quark sea of that era. As time went on, and the quarks condensed into protons and neutrons, the foundations of their existence would cease. Their day would be done. By their standards, that fraction of a second might be an eternity because time is really measured in events. Quarks were colliding with each other and transforming at a rate that is trillions upon trillions of times faster than the rate that chemicals interact with each other in a liquid. Therefore the scale of time would be completely different. There might be enough time for this form of life to evolve intelligence and have the realization that it was coming to an end. Having evolved for a whole millionth of a second, through innumerable generations of life and death struggles, in the end they had to face oblivion. What a bitter fate. How can a creature whose being has been shaped by constant struggle against death accept this? What could be done to somehow make continuing value out of those hard-won talents and wisdom?

Perhaps the only legacy available to this doomed form of life was to shape the universe that was to follow into a form that could produce us. Maybe they made some black holes to initiate galaxy formation in a universe that was otherwise too smooth. If the universe pulsates between Big Bangs and Big Crunches every fifty billion years (the Phoenix universe hypothesis) then such a project may be the only legacy available to the descendants of the life forms that occupy our cycle of the universe. Perhaps the form of the next Big Bang could be shaped by the beings that evolved out of the previous. In this wildest of speculations, the matter and the psyche of the universe would evolve through many cycles. Matter would turn into mind and mind back into the shape of the next round of matter.

Having gone totally off the speculative deep end, the existence of extraterrestrial life based on conventional chemistry should not seem outlandish. Lucretius and Bruno didn't think so. Early probes of Mars designed to detect life have found nothing conclusive, but perhaps the latest round will detect bacteria. The next best candidates are Europa and Titan (moons of Jupiter and Saturn respectively). Europa has a deep, ice-covered briny ocean. Titan has an atmosphere, an ocean of hydrocarbons (methane mainly), and a continent as well. But it will be a long time before anyone will land a probe there. After Titan, we will have to look into other solar systems. But the distance between stars is so vast that there is no way to get a probe to the nearest one in under fifty thousand years with today's best rockets, let alone bring it back. A hundred years from now, a rocket based on nuclear fusion might make the trip in less than a century.

If we can't go there, the only way to detect life is for it to come here or to communicate with us. But this wouldn't merely be detecting life, it would be detecting intelligent, technically advanced life. In the evolution of life from microbes to humans there was a long list of crucial developments analogous to the stepping-stones that our universe had to have in place before the biological era could even begin. If we had taken 25% longer to evolve through the necessary developments, then our sun would have gotten too hot before we came out of the trees. So we can't assume that creatures exist on nearby stars whom are sufficiently advanced enough for us to communicate with them or for them to come here any time soon.

But just as there must be life out there, there must also be advanced life. Some fraction of the stars that have life-supporting planets will have ones with radios. Even if it is a small fraction there will still be a lot of them. Perhaps some fraction of them are interested in contacting other life forms and are sending radio signals into space to do so. Based on this line of thought, numerous projects have been started to listen for radio signals from other stars. These projects are called SETI, the Search for Extraterrestrial Intelligence.

Clearly there are many opinions about the scientific value of the SETI projects. Not because people are "so what" about detecting a signal, but because there is no way to estimate the chances of success for a finite expenditure of money in a reasonable amount of time. Unless an alien society is pointing a very large radio telescope directly at us and deliberately broadcasting, and we have ours pointed at them, there is no chance of a successful communication. They would have to accidentally detect a radar signal beamed at them before they would have any interest in transmitting such a signal. Such radar signals have been propagating into space for only half a century; there are only about a hundred stars close enough to have potentially heard them and to have responded by now.

Despite the low probability of success, I think that SETI is a worthwhile pursuit, because the payoff would be so large by comparison to the investment in the cost of a modest program of listening. Tens of billions of dollars have been spent on probes of the Moon and Mars. In part, the effort was driven by the hope of finding microbial life there. The most recent important discovery, by the rover called Opportunity, was that of salty minerals that prove the past existence of liquid water there. If a SETI signal were detected it would have tremendous importance to our understanding of the evolution of intelligent life. It would give us the faith that dog-eat-dog natural selection is not fated to yield an evolutionary nuclear dead end whenever a creature takes up the use of high technology. It would quantify how probable the evolution of intelligent life is. It would give us a view into a completely new and fascinating ecosystem. It could fill in missing gaps in our history going back millions of years if these E.T.s had ever visited us before. But most important, it would hitch us into an intergalactic collective mind that has advanced philosophy, science, and technology to heights that we can't even dream of. Just how improbable this universe really is would be better understood by a civilization that had been thinking about it for millions of years.

Any E.T. that sent us a message would have to be way ahead of us. If they were behind by more than half a century they would not have the technology to communicate over interstellar distances. The chances of making contact with a civilization that is right at the dawn of the radio age (like ourselves) is miniscule. Their line of evolution would be billions of years out of step with ours based on statistics alone. Star formation in our vicinity has been going on for almost fourteen billion years. Any randomly selected center of life has a more or less equal probability of having started anywhere in that span of the last ten billion years. Like us they probably needed four billion years to evolve. The chance that it is within sync to a hundred million years is only one percent. Therefore it can be said with 99% certainty that any extraterrestrial contacting us would be at least a hundred million years ahead of us.

Think of the excitement of listening on the radio and getting a signal from a civilization that is in all probability millions of years more advanced down the scientific road than ourselves. Think of how scary that is, too. If push came to shove, we would be totally at their mercy! Should we answer their signal? Would they be dumb enough to send an advertisement of their presence to such adolescents? What would be in it for them?

Is Darth Vadar going to call? Probably not, because his civilization has probably destroyed itself with nuclear holocaust long ago. The fact that any E.T. had to learn to live with the bomb for millions of years is an important measure of their civilization. While creation through natural selection risks a universe dominated by monsters, there appears to be a kind of filter to weed them out.

Living with the bomb for millions or billions of years would be an achievement. With a great deal of cultural evolution, we have a shot at it. The positive changes that occurred internationally during forty years of eyeball to eyeball cold war gives us some reason to be optimistic. As we were under the gun, we saw the necessity for reform. But the proliferation of nuclear arms into less developed regions is a cause for great concern.

As concerned as we are about the existence of nuclear weapons, it is nonetheless a reason for comfort in assessing the psyche of an E.T. Along with nuclear technology goes the technology necessary for interstellar travel. If there has been an advanced E.T. in our neighborhood for the last million years, then they would have had the opportunity to make the multi-hundred year trip to our planet to see what was here many times. The trip would be so arduous that they would not do it very often however; we may have been apes the last time they took a look. They would have had the opportunity to destroy the planet long before anyone could have recorded the atrocity. The fact that we are here either confirms the benign nature of these neighbors or it indicates that they do not exist in our region of the galaxy. If they are within commuting distance (i.e. hundreds of light years) they haven't chosen to destroy us. Therefore it might be safe to answer the phone. But who should do the talking?

As fearful as we might be of a civilization that is a hundred million years ahead of ours, they have greater reason to fear one that is not spiritually advanced. They have learned to live with the power that technology confers and we have not. If we acquire their technology too rapidly by contact with them we could be the kind of threat that the Mongols were to ancient China. They probably won't contact us until we grow up.

An E.T. also might have second thoughts about opening a dialog with us, for in doing so they would be taking responsibility for our subsequent evolution. It was believed in ancient China that if you save someone from suicide, then you are responsible for them from that point on. A similar principle might apply here. An E.T. might not want to become part of our highly questionable Karma. Perhaps in a few thousand years we will have got our act together sufficiently to alter this assessment.

A related reason for either party to be coy about communication is that evolution and growth is something that only we can accomplish for ourselves. Direct contact with another being could stunt our growth; we could become like the child that never leaves home. We might be perpetually dependent and a sad shadow of what we might have become. The excitement of uncovering obscure layers of science would be taken from us. The need to search for deeper meaning in life would be smothered by E.T. gurus with intergalactic reputations. We would be like the medieval scholars who deferred to Aristotle on everything.

With so many pitfalls associated with contact, the question arises as to why they should do it at all. The only reason an E.T. would want to contact us is out of curiosity and an enhanced appreciation of the creativity of the universe. We like watching nature programs about all sorts of very different creatures, from the depths of the ocean to the wilds of a jungle in Africa. Most of us have evolved beyond exploiting these animals for the meager economic benefits they might provide. We are content with the psychic benefit of knowing that there is yet another way to live, and that they find their lives sufficiently meaningful to plough on through harsh conditions. In other words, the theater of our lives might be the principle value that we would provide an alien creature, a drama that might be spoiled by intervention.

So, it is possible that a neighboring E.T. would avoid contact with us, at least initially. However for drama and zoological curiosity they might make stealthy visits to this planet to observe us and our ways. But, because a round trip would be expensive and take hundreds of years, it is not likely that E.T. visits would be frequent. Our nearest neighbors may think that we are still in the Stone Age. The explosion of science and technology that has occurred in the last several hundred years might have been unanticipated. They may not have even begun to receive radio broadcasts first made seventy years ago. When they do start receiving television stations, what will their impression of us be? Fortunately, they probably will not be subjected to the twenty thousand murders that the typical kid sees on TV by the time he reaches adulthood, because the unfocussed signals will probably be too weak to decode at a distance of a hundred light years even by very advanced technology. Their first indication that something new is going on here might be from cold war radar beams that occasionally hit their vicinity.

So the answer to the question "Are we alone?" is undoubtedly "No!" But don't hold your breath expecting company. Before an

E.T. will be willing to hazard contact, it will have to see how we respect other living creatures to whom we are so closely related. They will also have to see how much respect we have for other humans as well. If we can't respect the lives of our close relatives, how would we possibly respect theirs? Undoubtedly they would also understand the need for us to mature on our own under the nuclear Sword of Damocles and not have the arrogance to interfere with such a deeply imbedded selective feature ("Galactic Selection", [3]) of life's evolution throughout this universe.

Chapter 13
Where Is the Signature?

The previous chapters have broadly described how the raw energy of the Big Bang has spontaneously evolved into all of the complexity and beauty that we see. In the latest phase, the pseudo life forms of the mind, including individual and collective psyche, evolved on the foundation of the biosphere. In the next phase, the various centers of psychic life in the universe will connect to form a universal collectivity of mind. We can only vaguely perceive this in the mind's eye. Only our piece of it is visible at this time. In these last two phases, this incredibly creative universe evolves creators with the power to transform the universe itself.

Was this improbable and incredible process the result of a trillion trillion throws of the universe-creating dice? Or was it all designed from the start in such a clever way that the subsequent evolutionary developments were, in a general sense, inevitable? For example, it is almost inevitable that Nazi-like civilizations will destroy themselves in this universe before they have the opportunity to destroy neighboring life forms. Is there a creator, or a team of creators, that occupies a larger universe of which ours is a subset? What might the motivation to create have been? Why didn't the artist sign the canvas? Is there a signature?

Answers to these questions can take many directions. As pointed out in the first chapter, the issue of whether or not the universe was created cannot be answered in a scientific sense because the anthropic principle provides too much wiggle room. However, trillion-trillion-to-one odds gives us the right to be suspicious. This suspicion can be investigated by assuming each hypothesis to be true and then playing each out to explore its likely consequences.

The consequences of the hypothesis that the universe was not created will be further explored in the next chapter. As earlier pointed out, even if this hypothesis is assumed, the improbability of a universe that creates life, and the life of the mind, makes our universe inherently valuable. Despite the pain and suffering that are implied by creation through natural selection, life is worth living. Life is inherently valuable. It does not need a creator to be justified.

Even if the assumption that there is a creator is not absolutely necessary for believing in the value of life, and for living a productive, adaptive life, it is still a more comfortable belief to possess for most people. There are numerous reasons for believing in a creator that are based on what might be broadly described as rational self interest. What is known as Descartes' square describes just one of the many varieties of this kind of motivation. It is summarized in Figure 13.1 below.

	THERE IS A GOD	THERE IS NO GOD
I BELIEVE IN GOD	Everyone is happy. Life has value. Moral order is maintained. God is content.	So what. Believing kept us from eating each other.
I DON'T BELIEVE IN GOD	God is angry and will damn. Individual freedom from moral restraint ultimately degrades the quality of life for all	We eat each other with impunity until nuclear weapons select us out.

Figure 13.1 Descartes's Square modified to include potential social consequences of belief in God or its lack.

In the original version of this analysis, the social consequences were not included. Only God's wrath or favor appeared in the boxes. Based exclusively on that criteria, it was clearly in one's interest to believe and thus avoid the unsavory possibility of dealing with an angry God. If that belief were wrong, the only negative consequence would be that one's individual freedom to be predatory towards other humans may have been needlessly inhibited by fear of a damnation that did not in fact exist (except as a symbol of the suffering that often rebounds on the perpetrator).

In some sense the social consequences that I have added mirror the heaven/hell consequences that Descartes and his generation believed would issue from a God who demanded belief. The damnation issuing from an angry God can be viewed as a symbol of the social chaos that can result in an unscrupulous society. For many people, the fear of damnation is a powerful motivator of civilized behavior. Also, the belief in rewards in heaven mute the pain of a life under stress. It can also motivate heroic behavior needed in times of crisis and war. This line of thought is summed up by the story of the family in therapy whose relative, Harry, thought he was a chicken. When asked why they don't simply tell Harry that he is not in fact a chicken, the reply was that they "needed the eggs."

Whether or not we "need the eggs" does not constitute evidence for a creator. Similarly, fear of an angry God is not a good reason for faking belief. The fear that not believing will incur wrath is propagated by religious institutions that are responding to forces of institutional natural selection. The religions that implant this fear acquire and hold members more reliably than the ones that do not. Therefore, the belief that God demands faith is common despite the fact that the evidence for a creator is not obvious. Where is the signature? If God is interested in our belief, the universe might have been designed in a way that made the act of creation more obvious. Therefore the assumption that the creator does not care about belief seems to be more supportable than the more commonly held belief that unquestioning faith is of overriding importance to the creator.

If the creator is disinterested in our faith, then Descartes' conclusion that it is better to believe is undermined to some extent. One might still hedge against the possibility of the wrathful type of creator. However, the absurdity of the situation in which humans would be punished for not believing, despite the absence of an obvious signature, is pointed out by a discussion between the famous mathematician and atheist Bertrand Russell and a reporter. The reporter asked him what he would say if upon his death he found himself at the Pearly Gates and God asked him: "Bertrand, why didn't you believe in me?" Bertrand said that his comeback would be: "Why didn't you provide me with more evidence?" I think the answer would be: "I didn't want to."

Even if the creator does not want to be apparent, we can still consider the deistic hypothesis. So let's address the questions that arise from the assumption that the universe was deliberately created. The question at the top of the list is: What can we know about the creator from creation itself? If the universe was created, then the creator had to be a fantastic artist/engineer/philosopher/everything. So much had to be just right that many have called it the Goldilocks Universe. Such an achievement must have been a supreme act of creativity. The creator must have been fascinated with the task and derived satisfaction from the results. How could a creator keep it a secret? Wouldn't a signature on the canvas be compulsory? Why would a creator resist the temptation to sign?

One common answer to this line of questioning, according to our established religions, is that the lack of an obvious signature is a test of faith. Faith in the tenets of these religions is an institutional imperative. As mentioned above, such faith is necessary for their survival. It is their psychic heritage, in the same way that DNA is our biological heritage. It is also an easy answer to the "why?" questions. It is not my favorite answer to the signature question or to the other why questions.

Perhaps a better answer is to observe that the signature paradox is nested in a somewhat silly anthropomorphism. Just because a human might feel compelled to claim recognition for the fantastic achievement of creating such a creative universe is no reason to assume that the creator would feel this way. To assume that the creator of a universe would need the ego boost that an artist gets from signing the canvas is to underrate the creator. Assuming that the creator wishes to be worshipped and stroked for the achievement is similarly an anthropomorphic, yet common, belief. On the other hand, we might do so for our own edification. Another reason to thank the creator is that it may be impolite not to do so. What if the creator does in fact have such sensibilities? But again, the absence of an obvious signature speaks against this possibility (without ruling it out).

But is an obvious signature really absent? Maybe we are just too insensitive to see it. Even before the age of science, many poets, artists, and prophets clearly saw the hand of a creator in the incredible reality that was before their faces. Now, with the eyes of science, we can see just how complex and improbable it all is at deeper levels than before. Are we being like a stubborn child who refuses to admit to a reality when we say that the tiny probability of getting such a universe from chance alone isn't scientific proof of the hand of a Master Watchmaker? Just because there is no mathematically rigorous proof does not mean that there is no evidence. Each of the fourteen stepping-stones that helped get us here are individually improbable. However, each taken alone might be described as circumstantial evidence in a court of law. When the fourteen are taken together, the case would be solid except for the loop hole that is provided by the anthropic principle.

One line of investigation that will be more possible in the future is to theoretically determine just how fertile and improbable this universe is in relationship to other possible universes. If our universe was found to be hyper-fertile, in relationship to other more probable but less fertile candidates, then we would be justified in suspecting the hand of the Master Watchmaker. It would then appear to be fine-tuned for life in relationship to these less hospitable and more probable places. If the process for producing universes were random, then an intelligent creature might be more likely to find itself in one of these more numerous types of marginally fertile universes. If greater scientific knowledge, or the discovery of numerous E.T.'s, enables us to call our universe super-fertile, then we might have caught the creator in the act.

It will be a long time before our science can yield anything definitive on this issue. Perhaps the science of an E.T. is good enough to comment on whether or not we live in a super-fertile universe. The discovery that E.T.'s are numerous would be an important data point for this study. If they are numerous, then they are also benign. This follows because they would be nearby and have had the opportunity to abuse us long ago if it was in their nature to do so. But as was pointed out in the last chapter, we may not want to hazard contact just for the chance that we might get some insight into this line of thought.

Even if we cannot be completely scientific about the super-fertility of this universe, we can still make some preliminary observations.

The mere fact that this universe has a trillion trillion stars implies that it is very fertile. Even if the odds are only 1% that life will evolve on any of them, then there are still a billion trillion centers of life here. The anthropic principle needs only one. Solar systems are perfectly ordinary in this universe. Our sun is a very typical star with a very typical ten-billion-year life expectancy. Therefore, the environments in which life can form must be common.

In addition to providing the chemical raw materials for life, and planetary homes in abundance, this universe provides many sources of long-term energy as well. In a universe that could just barely produce life, one would expect only one source of energy. Ours has at least four: fusion energy from stars; the beta decay of long-lived potassium40 that is the strongest contributor to the heat of the Earth; long-lived alpha decay from heavy elements like uranium and thorium; and tidal gravitational energy like that which keeps the moon of Jupiter called Io in a molten state. Any one of these sources could conceivably serve as the food for a life form. The fact that we have them all appears to go beyond the minimum requirements of the anthropic principle for a long-lived energy source. However, it is hard to see how complex creatures like ourselves could have come to be without the prodigious energy from the sun.

Even though this is a very improbable universe, life within it must be very probable. The chemical materials, the environments, and the energy sources for life are common. When these things are present, life wastes little time in getting started. After it starts, there is enough environmental instability to flog it up the evolutionary ladder before these conditions disappear. This universe must be teaming with intelligent life. If this assessment is born out by facts in the future (e.g. contact with E.T.s) it would then meet part of the criteria for a designed, super-fertile universe. This would give one pause about whether or not it was created. Even then, the jury must remain out until a theoretical assessment of the other possible, less fertile universes can be made with scientific confidence. If this assessment indicates that most life forms could expect to find themselves in the more common, marginally fertile universes, then a case could be made for deliberate design.

In the meantime, our Sherlock Holmes-type snooping around can go in other directions. New evidence might be based on observing some sort of built-in bias in the structure of the universe. If it is not a signature, then perhaps it is a signpost. This signpost would reflect some kind of preference or ulterior motive on the part of a creator. The special feature that would reflect such a bias would have to go beyond the requirement that the universe be fertile. This is because any physical characteristic that is absolutely needed for life can be dismissed by the anthropic principle. The structural anomaly that is needed here must also be inherently improbable in order for it to be used as circumstantial evidence for the act of creation. The universe should be able to produce life without it and it should be unlikely as well. It should also have a profound impact on the course of the evolution of mind.

Is the existence of fissionable materials in this universe such a signpost? Does it say, "Go this way and not that way?" Is it necessary for life to evolve? Is it likely to be in any universe merely by accident? In my view the existence of semi-stable uranium and thorium is very suspicious. It looks a lot like a signpost. It certainly tells us something about how we can expect to live a successful life in this universe. If we don't pay attention, humanity will blow its chance and some other animal will have a shot at being the way of the future (e.g. Admiral Richover's smart rabbit). This gives the idea that the meek will inherit the Earth new meaning. It is clear that if some kind of Nazi makes a play for the future of humanity there will be nothing to inherit until the ecosystem can recover thousands of years later. The warning that is written on this sign post is stark. To a creator, that has chosen natural selection as a tool for creating intelligent life, this signpost probably seemed necessary.

Naturally occurring fissionable elements do not appear to be necessary for life. If they were necessary, then it might just be another aspect of the anthropic principle. These elements provide about 40% of the heat of the earth at this time [1]. As described in Chapter 5, geothermal energy is necessary for the stability of the oceans. However, Earth could have maintained an active interior on the basis of the decay of non-fissionable potassium40 and residual heat alone. Potassium40 was the majority player in the geologically recent past. At this time it still contributes more heat than any other source. Therefore, the fissionable stuff does not appear to be a necessary accident. Only bad luck or deliberate design accounts for its existence in our reality.

If it was an unnecessary nuclear accident, then how improbable was it? How much does the structure of a universe have to be modified to get rid of these elements? The answer is not much. Uranium and thorium are both unstable elements. If the decay rate of these elements were one hundred times faster than it is, then there would be no natural uranium or thorium to speak of. For all practical purposes, we would live in a Nuclear Weapons Free Universe if we did not have long lived uranium and thorium in the ground.

But is a fertile universe that is also nuclear-free possible? How different would such an ideal universe have to be from our own? Examination of the relationship between the decay rate of heavy elements and the charge on the electron shows that this alternate universe would differ from our own by 1/3% on the charge of the electron [2, and see Appendix I]. If the fundamental unit of electric charge were increased by this tinny amount, then all of the elements with ninety or more protons in their nucleus would decay too rapidly for them to exist in nature for the five billion years that it took for us to evolve from the solar nebula.

How much can the charge on the electron increase before some other element becomes fissionable? The answer is about 8%. For this increase, all of the elements above platinum (e.g. gold, mercury, thallium, lead, and bismuth) would become unstable as well. Only platinum 198 would survive and it would be at a point of fission instability that is comparable to where uranium is in our universe. However, an increase in the electron charge of this magnitude would alter the burning of stars significantly. In order to burn, stars would have to be larger than our sun. However, life bearing solar systems probably would still exist and some of their stars would be long lived. At the ends of their lives the giant ones would still convert helium into carbon but at twice the temperature that is needed in our universe.

Therefore, all of the possible universes with electron charge ranging from 1/3% to 8% more are Nuclear Weapons Free Universes. These possible universes probably also contain sentient beings. They would have long lived stars that produce the elements of life (but in different abundance). However the self reflective creatures of these universes would not be constrained to curb the violent ways that billions of years of natural selection had bred into their genetic being.

The next natural question is: how much can the charge on the electron be decreased and still get an intelligence-bearing universe? One limitation comes from the helium2 resonance that is involved with the burning of hydrogen into deuterium and helium. This resonance would allow a decrease of about 1%. For this change, helium2 would become stable and the rate of hydrogen burning would go up by millions. Stellar end of life would occur too soon to get the evolution of intelligence at our level. However, one caveat on this is the possibility that life could exist on geothermal energy in these universes with short-lived stars.

The observation that decreasing the electron charge increases the abundance of fissionable material leads to the realization that in these alternate universes almost anyone can have the bomb. If the electron charge is decreased by more than 1/6%, then the fissionable variety of uranium ($Ur235$) becomes stable enough to exist in nature at almost a bomb-grade concentration. So, for all universes with less than this much charge on the electron, control of nuclear weapons would be much more difficult. Building reactors would be much easier. Ore deposits of uranium would routinely make natural nuclear reactors that would point primitive and vicious creatures in the direction of nuclear holocaust. In fact, a natural nuclear reactor formed 2 billion years ago in what is now the Republic of Gabon in Africa.

In the Uncontrollable Nuclear Weapons Universes, civilizations, countries, or individuals could obtain fissionable materials and build bombs with far less knowledge, sophistication, and resources than are required in our universe. It is highly doubtful that a technically intelligent civilization would be able to stay in that state for very long. Any mad and rich being would have the power to bring it all down. A universe could not have an electron charge less than 99.85% of our value and retain intelligence for very long. Cognition would be a flash in the pan that ended in mushroom clouds. However, the anthropic principle would not exclude this kind of universe from the group that contained cognition long enough to evaluate the improbability of it.

Our Improbable Universe

From the above discussion, the kind of universe in which control of nuclear weapons is possible occupies a full range of electron charge of about 1/2%. One-sixth of a percent below the present value, the life expectancy of technical civilizations is optimistically numbered in centuries. This is because anyone can get the bomb in these Uncontrollable Nuclear Weapons Universes. One-third of a percent above the present value for the electron charge are Nuclear Weapons-Free Universes. In these universes the creatures are free to pursue the philosophy of domination through military conquest forever. Within a very large range of charges (e.g. about 8% more), stars behave close enough to our own that intelligent life is feasible. The odds that our universe should randomly fall in the range that produces a Controllable Nuclear Weapons Universe are only about one in twenty. This situation is summarized in Figure 13.2 below.

The bottom line is that if you could visit all of the alternate universes, there would be less than a 6% chance of finding intelligent life that possessed our kind of international reality. This reality makes self-destruction possible but not inevitable. Therefore, the anthropic principle would predict a preponderance of long-lived Nuclear-Free Civilizations. Amongst the collection of all centers of cognition, our kind is in the minority. Nuclear Free Universes are in fact much more probable than the one we have. There are only two explanations for the electric charge that we have. Either it was what at first sight appears to be very bad luck for us but good luck for our neighbors on nearby stars (because we must learn to live with the bomb before we can get to them). Or it was a signpost that was stuck in the ground by a creator who wanted to point out a one-way path into the future. That one-way path is the path of peace. The one of war leads to oblivion. This is true whether or not there was a creator.

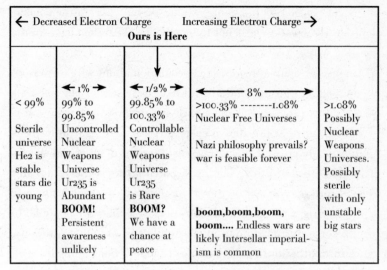

| ← Decreased Electron Charge | | Increasing Electron Charge → | | |
| | | **Ours is Here** | | |

< 99%	← 1% → 99% to 99.85%	← 1/2% → 99.85% to 100.33%	←————— 8% —————→ >100.33% --------1.08%	>1.08%
Sterile universe He2 is stable stars die young	Uncontrolled Nuclear Weapons Universe Ur235 is Abundant **BOOM!** Persistent awareness unlikely	Controllable Nuclear Weapons Universe Ur235 is Rare **BOOM?** We have a chance at peace	Nuclear Free Universes Nazi philosophy prevails? war is feasible forever **boom,boom,boom, boom....** Endless wars are likely Intersellar imperialism is common	Possibly Nuclear Weapons Universes. Possibly sterile with only unstable big stars

Figure 13.2 Esitmated physical, biological and political consequences of different values of the electron charge.

Whether or not the existence of uranium and thorium is a signpost is still open to debate. Even at the level of 94% confidence, there is still room for doubt and uncertainty. It is still not obvious because of the subtlety of the implications outlined in the Table above. It took most of this chapter to scratch the surface of this issue. Therefore, we are still left with the search for an obvious signature. Having looked under many stepping-stones, we have to consider the possibility that it is simply not there. A glib explanation might be that there was no creator to sign. But still, we are considering the deistic hypothesis here. Another glib explanation is that perhaps the creator wasn't too proud of the results and did not want to be blamed for the outcome. George Sand has said that she would rather not believe in a God that was indifferent to our sufferings. Even Genesis documents the despair of the creator in the Old Testament when Adam and Eve claim they weren't misbehaving. After many rounds of the cat chasing its tail game, we are driven back to the conclusion that the creator chose not to be obvious. This then leads to the question why.

Among the many directions of speculation about why there is no obvious signature is the one which I like the best. This answer rests on the observation that if there is divine purpose in creation, it is to be as creative as possible in every possible way. This universe is a fantastically creative place. The vast number of stars and species of beetles attests to this. The huge potential for variety that this universe offers must not be muted in any way. We have a universe in which three varieties of conscious psychic beings are possible. In this universe it is possible to intellectually sustain either a theistic, agnostic, or atheistic approach to life. The kinds of meaning that can be found in life by all three types of beings are found in this most creative universe. If the canvas had been obviously signed, then we would have a universe in which there were only theists or deists. The universe would then contain only their answer to the meaning of life. This answer simply says that the creator thought it was meaningful and that's the end of it. If the creator is also pondering the meaning of life, then a signature would deny the wisdom that can be found by agnostics and atheists. The signature would snuff the question from the universe's reality. The added meaning to life that the created could provide for the creator would be greatly diminished. The meaning that the created could supply to the creator would be primarily derivative. It would have to come from the creator in the first place. How could the creator learn anything from it?

So a signature would make the universe a lesser place in many ways. The fact that it would be a lesser place is witnessed by the behavior of the authorities in the theocratic states that have existed. Diversity of opinion is suppressed and only one kind of human being is considered to be acceptable to God. Consider the authority that has been wheeled in such states. Then consider how much more dogmatic and uncompromising power would be if those authorities had an obvious signature to point to. The psychic monoculture that would result is anathema to the call for diversity that the universe is screaming at us. It is not what a creator of this universe would want, so the canvas remains unsigned.

An answer to the signature question that is related to the preceding suggestion and to a reason that an extraterrestrial would maintain a hands-off policy is the following: only humanity can create the evolutionary changes in our individual and collective spirit that are necessary for us to produce a sustainable civilization in the long run. Any intervention of any sort would result in a dependency mentality that would derail the process. If there were a signature, we would have someone to depend on and someone to blame. As it is, we are stuck with ourselves. If we don't do it, we have no expectation that anyone else will. We either grow up or we take ourselves out of the running and some other species gets its chance. If the universe was created, then we better accept this tough love, and get on with the creativity of life.

Chapter 14
Value Added

We explored some of the implications of deliberate creation in the previous chapter. Now the hypothesis that our universe was a random happening needs to be considered more deeply. If the physical structure of this particular universe resulted from a mindless random process, then statistics imply that parallel, infertile universes must exist or have existed in great abundance. The odds of getting a universe with intelligent beings from a single throw of the dice is just too small, less than parts per trillion-trillions. Forty years ago, John Wheeler was fond of the concept of a pulsating universe. He proposed that only the total energy and angular momentum (rotation) would be carried over from one pulsation to the next. He hypothesized that everything else would be different. The strengths of the four forces, the number of elementary particles and their characteristics (e.g. charge, masses, spin, etc.), and details of the expansion would all be different. After trillions of pulsations, the structures that are necessary for an intelligent universe might arise by pure chance. This is a sequential hypothesis that keeps recycling the raw energy of the Big Bangs until witnesses arise to ponder the extreme improbability of intelligence.

This hypothesis relates to the anthropic principle, but in some sense it begs the question of cause as much as the hypothesis of deliberate creation does. One is forced to have faith that a mechanism for changing the physical parameters of the universe exists.

This is necessary if each pulsation is to be different than the previous one. How do the dice turn over? Until recently there was very little in the way of an explanation for this in the scientific community. The best ideas at this time are motivated by elementary particle theories of "Gage Fields" with "Spontaneous Symmetry Breaking." The symmetries can be broken in many ways just as a pencil that is standing on its point can fall in any direction. Each direction corresponds to a different combination of physical properties for an evolving universe.

A theory based on spontaneous symmetry breaking in a system of many *scalar* fields has been proposed by Andrei Linde. It allows for many universes with different physical laws [1]. This new hypothesis proposes an extremely large metauniverse, in which smaller universes pop into existence in parallel and sequentially (see Fig 2.2). This larger metauniverse may be a 1,000,..., 000. (a trillion zeros) times larger than our puny sub-universe. The number of sub-universes could easily meet the requirements for producing many universes with intelligent beings purely by chance. In addition to many sub-universes that exist at different locations in the metauniverse, each sub-universe can spawned sub-sub-universes within themselves.

At last a mechanism for generating variability between universes has been proposed. This mechanism allows the relative strengths of the different forces and the masses of the elementary particles to vary from one sub-universe to the next. The vast majority of these sub-universes would be sterile. A very small fraction of them would be fertile. An even smaller number would contain intelligence because of the proper configuration of things like our stepping-stones. Of the intelligent sub-universes: eighty-five percent would be Nuclear Weapons Free Universes; about five percent would be

Controllable Nuclear Weapons Universes like our own; and about ten percent would be Uncontrollable Nuclear Weapons Universes (see Table 13.2). The intelligent beings in these Uncontrollable Nuclear Weapons Universes would be almost certainly doomed to destroying themselves by the time they could consider these issues. On the other hand the intelligent beings in the Nuclear Free Universes are doomed to war without end. Was humanity unlucky or lucky to be in a Controllable Nuclear Universe? Only time will reveal the quality of our souls, which is the prime determinant of our ultimate fate.

Whether or not the inflationary theories proposed by Linde and others hold water in the long run remains to be seen. However, the observation that the symmetry breaking of two or more scalar fields can lead to vastly different universes rests on a strong foundation. To a great extent, this observation validates the anthropic principle's approach to explaining how our universe could spontaneously come into being. A creator might have chosen to create in this lazy way (all things could happen) or the metauniverse might have spontaneously precipitated from an unthinking physical reality. In either case, we are left to ponder the value and meaning of life on our own.

If there was no deliberate creation, or if the creator was so distant as to be effectively absent (this is the deist hypothesis), then it is up to us to figure out whether or not life has value, what it is, and how to create it. Left to our own devices in this way, we can't simply throw up our hands and say: "I don't know what life's value is, but I'm sure that the creator knows. So I'll have another beer and forget about it." As pointed out earlier, an apparent creator would automatically rob us of the need for dealing with this issue and the answers we might find.

The need to find value and meaning in life is more than an academic curiosity. The perception of value in existence is needed to create value and not to mindlessly destroy it. Seeing the value in life gives us respect for the lives of ourselves and others. A vision of the nature of true value is necessary to get through life's difficult and painful times. In some sense, its pursuit has been the central task of humanity going back to the primitive microbes from which we evolved. In some sense it is the pursuit of the entire metauniverse from which our sub-universe may have evolved.

In that huge metauniverse there would be an incredible predominance of sub-universes that never made it all the way through the Inflationary Era. They would oscillate on the brink of being and then recollapse back into nothingness. These universes that quickly recollapse, are more than nothing but not by much. Other universes would emerge from inflation with no CP Symmetry Violation. After a few seconds all of the matter and antimatter would mutually annihilate and they would become a lifeless ball of photons and neutrinos expanding forever. These sub-universes are more interesting than the ones that last less than a microsecond. They have everything that the still born ones do and more. They are more valuable in that they add a new flourish to the metauniverse that might not otherwise be there. They add structure and dynamics to a place that might otherwise be uniform and static. If there is anyone watching the fireworks display, they add a burst of light in the darkness. As our science improves we will get a more clear view of these other universes in the mind's eye.

Among the group of sub-universes that actually exist for a finite length of time there are those that have six quarks, CP Asymmetry, thermal nonequilibrium, and a reaction that changes quarks into electrons. These universes can produce an excess of long-lived matter over antimatter. They have light and neutrinos, but in addition they have long-lived protons and electrons. They have everything that the ball of light universes do and more. They are therefore more varied and of greater value. They have stars, atoms, and a shot at life.

Among the sub-universes that have matter, there are many that have a much stronger weak force or a strong force that is only 0.5% stronger. In these sub-universes, stars are short-lived because of their rapid production of helium2. They burn fiercely for awhile, but then go through their death throes in a matter of a few million years instead of billions of years. The sub-universes with long-lived stars have everything that the others do, but in addition they have the potential to produce intelligent life. They add value that goes above and beyond what the lesser universes have.

While stars would burn too quickly where the Strong Force stronger, they would fail to burn at all if it were weaker by more than a few percent (i.e. too little helium2 and beryllium8). In such sub-universes, stars would release gravitational energy for a few hundred million years and then either cool to a ball of gas like Jupiter if they were small, collapse into neutron stars if these were large, or collapse into a black hole if they were very large. Any form of life would either be blown to smithereens, if supernovae still occurred, or permanently freeze as the star went quietly to its grave. This would happen long before planets could evolve the intelligence that was necessary to get away from this fate.

Everyone of the fourteen stepping-stones discussed earlier can be examined in this way. The sub-universes that don't have them positioned in just the right place for intelligence to evolve are lesser places than our universe. Our universe has everything that these places do, but in addition it has us and our fellow travelers (i.e. bacteria, plants, animals, E.T.s, etc.). It adds value that the others don't have.

The microcosm of planet Earth also has had its stepping-stones. The sun is just right. The Earth's orbit is just right. The need for some stability was fulfilled by the just-right mass and orbit of Jupiter. And it is good to have a big Moon. The amount of instability has been just right to occasion the evolution of more advanced creatures from simple microbes. These microbes developed all of the incredible machinery of the cell long before they evolved into multi-celled organisms. They had great complexity and value added at their own level. But if they had not made this step toward even greater complexity, then they would face certain extinction in a billion years when the ocean would have boiled away.

Life, with ever greater complexity, evolved on top of the foundations provided by the previous changes. The amount of stability has been just right to let the fragile, multi-celled organisms refine their life styles and genes before the next great crisis hit. Each time, a biological stepping-stone specific to our microcosm had to be in place. Each new change preserved some of the creatures and value of the previous era while generating new realms of diversity, complexity, vulnerability, and value. In turn, the vulnerabilities motivated further evolutionary change. Each new level grew out of the value of the previous level and added new value to it.

The whole process of evolution went into high gear when some apes took up the use of tools and then learned to make them. People evolved human spirit and learned to talk and to communicate it on a psychic level as never before. A completely new reality of collective mind had evolved on top of the biological level. Within this new sphere of reality, new complexities evolved as each revolution built new forms of being and value on top of the old.

As humanity evolved on the biological and cultural level, we overcame many individual and collective crises. Life was incredibly hard for most of our history by comparison to the ease of modern life. Our ancestors ate carrion and feared being eaten by lions for millions of years on the savannas of Africa. The long-term cooling of the planet led to ice ages in which the climate was harsh almost everywhere. People somehow survived a population bottle-neck about 60,000 years ago that is evident in our reduced genetic diversity. Later we developed farming when the big game gave out. We created great civilizations based on this new economic foundation. Political structures then evolved a new form of national being. Though nations provided the internal peace and stability in which many new forms of cultural developments could evolve, they also produced wars on an unprecedented scale. In the twentieth century, the process snowballed to the point where existence itself was at stake. Somehow we survived the Cuban missile crisis [2]. This sobering experience has helped lead to the end of the cold war (for now) but we will still face the threat of nuclear proliferation.

Over-population and ecological crisis are imbedded in much of this history. These problems are here today as never before. All of these disasters and triumphs have made us into what we are today. Though great losses have occurred, and new vulnerabilities have arisen, the overall process has added complexity, value, and human happiness. But along with this exponential growth in the number of humans and the complex entities of our collective mental being, a serious imbalance has developed. We are crowding out all of the other biological creatures of this rare and beautiful planet. We are subtracting the value that they could add. In greed and selfishness we are even undermining the ecosystem that supports our own biological needs. We are not the only value that is added to this planet. A hundred billion people will not increase the value of humanity by sixteen times the value of the six billion that we have now. It will in fact degrade it well below where we are today by destroying the value of the ecosystem. The quality of every human life would be vastly degraded by such over-crowding. The collapse of the ecosystem could even lead to the kind of economic and social collapse that, according to Captain Cook, turned the sculptors of Easter Island into cannibals. We, and all the creatures of this planet, are added value. It could have been a lifeless rock. If we aren't careful, we will *make* it into a lifeless rock.

If we can perceive the values created by this universe and ourselves, then we will go with the flow and be part of the creative process rather than a drag on it. We can learn to use our creativity to enhance, rather than to detract, from the values of planet Earth. We don't have to have a planet with one hundred billion people and no natural ecosystems. The value of human life is not increased by overpopulating the planet. It is certainly decreased by obliterating all natural beauty. The value that humans can create is increased by improving the creative qualities and the beauty in the lives of a sustainable population. We should use our creativity to evolve the spirit of humanity into the caretaker of this incredible planet rather than to become its scourge. We should look to the creativity of this rare and beautiful universe and say, "Yes! I want to be part of it and add to its value!"

Chapter 15
Our Valuable Universe

Comparison of our universe to other potential universes leads to the conclusion that this is a remarkably rare and creative place. The evolving creative processes started with the first moment of the Big Bang and continue today. At least fourteen physics-based stepping-stones and countless biological ones had to be traversed for the seed of raw energy to have grown into us. With the invention of complex language, we became creatures of a collective mind as much as creatures of a collective gene pool. Within the last several thousand years people have created social organizations of unprecedented scope. Civilizations have arisen and pushed the development of science and technology at an ever-quickening pace. The invention of printing, radio, television, and computers threw oil on the fire, affecting rich and poor. We are now part of a worldwide collective psyche that is shaping all of us and is shaped by us all. We are it and it is us. Humanity has become one as never before.

The power of this collectivity is incredible. For the first time in human history, middle-class people can have a quality of physical life that exceeds that of a Syrian king—a mere three thousand years ago—and be free of his diseases and assassins as well. Just about anything that can be imagined can be made to happen through collective effort and modern technology. Going to the Moon was a pipe dream for centuries, yet it happened forty years ago. Modern computers make virtual realities that would turn any ancient magician green with envy. Simulations of complex weather systems and others can predict things that were previously in the domain of those who looked into crystal balls or piles of chicken bones.

As time goes on, the evolution of these capabilities will make almost any dream a potential reality. In this century, the development of fusion reactors will make cheap, safe, carbon-dioxide-free and pollution-free energy available. This will happen when and if we ever develop the necessary collective will. If the level of commitment that was applied to the Moon program were applied to building fusion reactors, they would have become a reality thirty years sooner.

The fact that we find it difficult to generate intelligent and appropriate collective commitments to the future is one symptom of the many worms that lurk in this apple. It could be so sweet but often it is not. The evolution of human spirit on an individual and on a collective level has yet to produce sustainable, intelligent life on either level. We are immature adolescents with a foolhardy intoxication with the present. Our lack of interest in the past and the future deprives us of seeing the creativity, the majesty, the beauty, and the frailty of the evolutionary process that we are all a part of. Without this vision we cannot participate as constructively as we might. In fact, this lack of vision often leads to incredibly stupid and ultimately self-destructive behaviors.

Our potential for turning our back on the work of the universe exists on all levels of human reality. On the political level, we can destroy the world, or the part of it that we love, with nuclear holocaust. The end of the cold war has given humanity a much-needed temporary reprieve from this doom. But unless we quickly evolve international politics to a more peaceful form of competition, and institute meaningful arms control, we will lose it all. It is only a matter of time. If the odds of a nuclear catastrophe is only 0.1% per year, then the expected lifetime of humanity is only a thousand years. This is only 0.00001% of the time that the Big Bang took to create us. We should and must do better than that! We do not deserve to call ourselves an intelligent life form if we cannot live up to the nuclear challenge that is planted in our earth.

On the biological level, we risk destroying the biosphere that was our mother in an even shorter amount of time. Greenhouse warming from the burning of fossil fuels could do for the ocean what our benign sun did not. If all of the people in the world start burning fuels at the same rate as first-world people, a run-away greenhouse effect could start within a couple of centuries. The locked-in heat would drive water from the ocean into the atmosphere. The extra water vapor would then lock in more heat and the whole planet would cook.

A runaway greenhouse effect is only one of the many threats that the runaway propagation of humanity might inflict on the biosphere. Far more likely is the continuation of the destruction of habitat that is so common today. If we do not curb our ways soon, the great biodiversity of this planet will be lost forever. The other creatures of this planet all reflect the incredible creativity of this universe. They have also paid heavy dues to evolve into the beauty that they are today. Trillions of life-and-death struggles have gone into the shaping of each species over millions of years. To kill the last viable population of a species is equivalent to annihilating all the trillions of ancestors that suffered pain and death to make it what it is. It destroys a unique and complex beauty that required a unique and complex universe to evolve.

Perhaps our greatest risks exist on the level of individual human spirit. The problem is greatest here because our understanding is less. This issue is pivotal because healthy and happy individuals are not likely to tolerate life-threatening situations for long. The solution to the other problems will probably be found if the individual is strong and happy. But in the midst of great material well being we seem lost at sea when it comes to creating real human happiness. Our fiercely individualistic, modern philosophy of life has failed to acknowledge that human beings of the past were always imbedded in a society of kin and tribesmen from whom they could get the love and respect that is fundamental to human happiness. The rugged individualism of modern life encourages people not to give love, so we don't get it. We are starved for love and respect and we don't see that it is something that we have to create in order for it to exist. Even the children are often deprived of this basic human need for love.

In a broad sense, humanity needs love in order to evolve into a sustainable long-term future. We need to love all the forms of beauty that this universe has evolved. Appreciation of the remarkable struggle that these other beings had to endure should deter us from mindlessly destroying them. Appreciation of the incredible improbability of a universe in which these things could happen adds to the value of their lives and our own. If we are to survive, the universe demands that we learn to love life and to see our lives as imbedded in a larger life process that must also be loved. If we can truly value this remarkable reality, we will not senselessly destroy it in a frenzy of political or ecological violence.

The view looking back across the eons of time—the innumerable stepping-stones, and the lucky chances of our biological history—is awe-inspiring. You have to be a true nihilist to say "so what?" when confronted by this panorama. Whether the foundations under this reality were put in place by a clever creator or whether it is by chance amongst a few fertile universes in a zillion, this is a fundamentally unique and valuable place and the life forms within it are its finest gems. Whether we look to a creator or to a creative universe we see the same imperative: protect this creation and enjoy the show.

Appendix
Long Lived Radioactivity

Radioactive nuclei decay by spontaneous fission, Beta decay (the emission of an electron or and antielectron), or Alpha decay. Alpha particles are just fast moving nuclei of helium4, two protons and two neutrons. They are bound into a larger nucleus by the nuclear force, but at the same time they are repelled by the electric force. For uranium and thorium, the associated nuclear binding energy is less than the electrical potential energy by 4 million electron volts of energy (the electrons in your TV picture tube strike the screen with about 15 thousand electron volts of energy to make it glow). Therefore these bound helium nuclei want to leak out but their exit is impeded at the nuclear surface (at radius $R = 8.7 \times 10^{-15}$ meters) where the nuclear force is stronger than the electric force and the particles bounce back into the heart of the nucleus. But for uranium238 and thorium232, one bounce out of 10^{38} does not happen and the helium nucleus "tunnels" through the barrier.

Heavy nuclei tend to decay in this way because they have so many charged protons in the nucleus. This number of protons is usually called Z. For uranium $Z = 92$. Its first decay product is the very unstable thorium 234, which goes on to decay by the alpha emission process in only twenty-four days. After a long sequence of decays, which include radium along the way, the nucleus ends up as very stable lead 206.

The tinny probability of tunneling for uranium238 is very dependent on the strength of the electrical repulsion force, which is proportional to the Z of the daughter nucleus (Z = 90 for thorium). The calculation for this probability is performed by R. Leighton in Eq(9) on page 527 of *Principles of Modern Physics* (McGraw-Hill, 1959). This equation relates the tunneling probability to the electric charge of electrons and protons. However, he does not get involved with the "what if question" of a different universe with a different electric force (electron charge). This calculation is done here for the sake of completeness and so that others can attempt to falsify the assertion that a 1/3% increase in electric charge would result in a 100-fold increase in the decay rate of uranium and thorium. From Eq 9 on p527 of Leighton, the natural log of the tunneling probability, T, is given by:

$$\ln(T) = 2.97(e'/e)(ZR)^{1/2} \qquad -3.95Z(e'/e)^2/E^{1/2} \qquad (A1\text{-}1)$$

Where Z is the number of protons in the daughter nucleus, $R = 8.7 =$ the daughter nuclear radius in units of 10^{-15} meters (ibid p 529, Eq 12), e is the electric charge we have in our universe, e' is the alternative universe electric charge, and E is the energy of the emitted alpha particle, in units of millions of electron volts (MeV), which also depends on e'/e.

The increase in E with e' can be estimated by noting that the kinetic energy of the alpha particle at an average position inside of the nucleus is relatively invariant in that it is determined by the standing wavelength there. It is straightforward to show that at a distance of half the nuclear radius from the center (of a uniform sphere of charge) the electric potential, E_{ep}, is 37.5 % higher than at the nuclear surface. The kinetic energy, E_k, of the alpha particle at this point will be fairly independent of e' so that the standing quantum wave will fit into the nucleus. This then is a good location to evaluate the dependence of E on e' (note that the end result will be fairly independent of this assumption). As long as the chosen point is not near the surface, the nuclear force potential, E_n, will be independent of location because once the alpha particle is surrounded by nucleons it experiences not net force. In traversing from this average point in the nucleus to a point far from the nucleus the alpha particle's remaining energy is then:

$$E = E_k + E_{ep} - E_n \qquad (A1\text{-}2)$$

But only the electric potential energy changes with the electric charge so the change in emitted energy is:

$$E' - E = \Delta E = \Delta E_{ep} = E'_{ep} - E_p \qquad (A1\text{-}3)$$

At the chosen half way point the electric potential in Joules is then (ibid, p 527):

$$E_{ep} = (1.375 \ @ \ \text{half way point}) \ (2e = \text{alpha charge})(eZ)/ 4\pi\epsilon_o R$$

$$E_{ep} = 2.75Ze^2/4\pi\epsilon_o R \qquad (A1\text{-}3)$$

Evaluating this for $Z = 90$, $e = 1.602 \times 10^{-19}$ Coul., $R = 8.7 \times 10^{-15}$m, and $\epsilon_o = 8.85 \times 10^{-12}$ Farads/m and converting to MeV (divide by $e \times 10^6$) gives $E_{ep} = 40.979$ MeV. Repeating the exercise for a 1/3% increase in e gives $E'_{ep} =$ gives $(e'/e)^2 E_{ep} = 41.253$ MeV. So $\Delta E = 0.274$ MeV. Therefore increasing the electric charge by 1/3% for uranium238 increases E from 4.18 MeV to $E' = 4.454$ MeV. Going back to the tunneling probability the difference in $\ln(T)$ for these two universes is (note $e'/e = 1$ in our universe and 1.00333 in the other with a 1/3% higher charge):

$$\ln(T) = 2.97(ZR)^{1/2} - 3.95Z/E^{1/2} = 83.106 - 173.88 = -90.775$$

and

$$\ln(T') = 2.97(1.00333)(ZR)^{1/2} - 3.95Z(1.00333)^2/(E')^{1/2}$$
$$\ln(T') = 83.383 - 169.57 = -86.19$$

and

$$\ln(T) - \ln(T') = \ln(T/T') = -4.59$$

So:

$$T/T' = e^{-4.59} = .01 \qquad\qquad (A1\text{-}4)$$

So a 1/3% increase in the fundamental charge in our universe would result in a decay rate increase for uranium238 of one hundred fold.

To test the sensitivity of this result to the point where the kinetic energy and momentum of the wave function is invariant with e'/e, consider the result if this point is chosen to be the center of the nucleus. Here E_{ep} is 50% higher than at the edge and 9.1% higher than at the point assumed above (half way to the edge). This then gives E_{ep} = 44.708 MeV. Repeating the exercise for a 1/3% increase in e gives E'_{ep} = 45.006 MeV. So ΔE = 0.298 MeV. Then $\ln(T'/T) = -5.02$ and T/T' = .0066. This is about a 34% reduction in the estimated lifetime reduction factor. Similarly if we evaluate E_{ep} at 71% of the nuclear radius it decreases by 9.1% and the uranium238 lifetime reduction factor (tunneling probability ratio for a **1/3% increase in electric charge**) would increase by 60% over the initial estimate above. So the estimate here, with error bars for this ambiguity, is **T/T' = .01 , +51%/−34%**.

A similar exercise for uranium235 and thorium232 (with suitable adjustment of R and alpha particle energy) gives lifetime reduction ratios of 0.014 and 0.008, respectively.

Based on the above numbers, TableAI-1 below gives the abundance reduction factors for these isotopes, for various changes in the electron charge, at a time that is 6 billion years after they were created in a supernova assuming equal abundance at that time.

Table AI-1 Calculated half lifetimes and abundance (6 billion years after formation) for uranium and thorium for a range of changes in the electron charge from $-1/3\%$ to $+2/3\%$.

% Change in charge	U238 % left	U235 %left	Th232 %left	U235/ U238 %	U238 Half life billion yrs	U235 Half life billion yrs	Th232 Half life billion yrs
0	39.68503	0.262871	74.29971	0.662394	4.5	0.7	14
0.1	2.511266	5.04E-08	28.18605	2.01E-06	1.128787	0.194258	3.284168
0.167	0.009087	1.13E-20	3.524499	1.24E-16	0.446899	0.082296	1.243153
0.333	7.29E-39	5E-183	7.47E-15	6.8E-143	0.045	0.0098	0.112
0.4	4.2E-100	0	2.49E-41	0	0.014816	0.004152	0.042395
0.5	0	0	2.4E-180	0	0.004469	0.001152	0.009945
0.667	0	0	0	0	0.000444	0.000135	0.000883
-0.1	79.30835	19.22861	93.26866	24.24537	17.93961	2.522414	59.68026
-0.33	99.04128	91.71918	99.75209	92.60702	431.7124	48.11367	1675.51-

In Beta decay, a nucleus emits an energetic electron or antielectron. Its sensitivity to the charge on the electron is less than that of Alpha Decay. For a $1/3\%$ increase in the electron charge, the lifetime of potassium40 and its abundance would increase slightly over its present abundance. This lifetime is determined by the strength of the weak force and the amount of energy associated with the decay raised to the fifth power (ibid Eq 23, p 538). The amount of energy is influenced by the electric charges of the nuclei involved. For potassium40 the radius is about $R = 4.8 \times 10^{-15}$ m (scaling from the uranium radius above as nucleon number to the $1/3$ power). The electric potential energy for an electron at this radius is:

$$E_{ep} = Ze^2/4\pi\varepsilon_0 R \text{ in Joules} \qquad (A1\text{-}5)$$

For $Z = 20$ (the daughter atomic number) this is 6 MeV. A change in the electron charge of $1/3\%$ would change this energy by 0.04 MeV. The energy release in the decay of potassium40 is 1.33

MeV. This would be decreased by 3% by increasing the electron charge by 1/3% . The decay rate would then decrease by a factor of $1.027^{-5} = 0.88$. So the half-life would increase from 1.3 to 1.5 billion years. The abundance and contribution to the Earth's heat generation would then be increased by 50% at the 6 billion year mark (e.g. after the last supernova). So the loss of uranium and thorium would decrease the heat generation by about 40%, but the gain from more potassium would offset 1/3 of this. Earth would end up with about 75% of the geothermal power that it presently has. Essentially none of it would be from fissionable elements.

Table AI-2 Calculated half life and for potassium40 and % remaining after 6 billion years and its contribution to geothermal power generation along with that of other nuclear species.

% Change in charge	Potas.40 % left	Potas.40 Half life billion yrs	Potas.40 GeoPower TeraWatts	U238 GeoPower TeraWatts	U235 GeoPower TeraWatts	Th232 GeoPower TeraWatts	Pu244 GeoPower TeraWatts	Nuclear % of 44TW Earth Now
0	4.079724	1.3	13.00893	10.00011	0.450933	10.00023	3.77194E-20	76.04591
0.1	4.615703	1.352172	14.15012	2.522733	3.11E-07	16.17185	2.26245E-78	74.64706
0.167	5.000259	1.388293	14.93019	0.023057	1.65E-19	5.342245	1.1067E-188	46.12613
0.333	6.043035	1.482	16.90289	1.84E-37	6.1E-181	1.26E-13	0	38.41567
0.4	6.500774	1.52159	17.71013	2.66E-98	0	1.11E-39	0	40.25029
0.5	7.223794	1.582654	18.92054	0	0	4.6E-178	0	43.00122
0.667	8.537802	1.690145	20.93998	0	0	0	0	47.59087
-0.1	3.588164	1.249841	11.90068	5.012984	9.153735	2.944801	0.000214605	65.9373
-0.33	2.618109	1.141698	9.505844	0.260143	2.289067	0.112183	9.73129718I	49.7694

Table AI-2 shows the calculated lifetime and abundance of potassium40 verses the change in electron charge. It also shows its estimated contribution to Earth's geothermal power generation along with that from uranium and thorium. This is based on an estimate that residual heat (note that this refers to the heat of cooling and that most of this was the result of potassium decay billions of years ago) is providing about 25% of the present geothermal total of 44 TeraWatts, potassium is providing about 35%, and uranium and thorium are doing the rest in equal measure (this equality based on their relative cosmic abundance of 1:3 and their lifetime ratios of 1:3). From the last column it can be seen that for the range of electron charges considered (–1/3% to +2/3%) the nuclear contribution to geothermal power does not fall by more than half of Earth's present total nuclear production of heat.

Decreasing the strength of the weak force by 6.6% would lengthen the life of potassium40 by 14% (from 1.3 billion years to 1.57 billion). That amount would equal the increase of lifetime that would result from a 1/3% increase in the electrons charge. This would leave us with twice the amount of geothermal power from potassium40. This would compensate the loss from the lack of uranium and thorium. As pointed out earlier, the strength of the weak interaction is important to the workings of stars and the ability of them to explode in supernovae. This percentage increase would allow small stars to burn for billions of years and it is unlikely that it would cause supernovae to fail.

A final point concerns the issue of whether or not fissionable materials are absolutely necessary for the production of nuclear weapons. Without fissionable materials, a fission bomb could not be made, but what about a fusion bomb? All fusion bombs use fission bombs as triggers, but is this necessary? It turns out that it is. The incredible temperatures necessary to initiate fusion can only be produced in a sizable sample with a fission bomb trigger. When a nucleus splits into two nuclei of nearly equal size, the process is called fission. Huge lasers may soon produce fusion at the center of a tinny pellet but initiation of a self-sustained fusion reaction (bomb) would require a laser that was bigger than a battle ship.

Fissionable nuclei can be bred from non-fissionable nuclei by adding a neutron in a nuclear reactor. In a universe in which no naturally occurring fissionable material exists it would be very hard to get going on this because there would be nothing to build the first breeder reactor with. However, it is conceivable that an alternate source of neutrons could be obtained from particle accelerators, or even reactors based on controlled nuclear fusion. Therefore, a truly Nuclear Free Universe requires a 1/3% increase in the electric charge in order to eliminate thorium. The eradication of uranium235, the only material from which bombs can be made directly, would occur for only a 0.1% increase in the electric charge. However, this would only postpone the time when civilization was forced to grow up. The 1/3% increase would give us a truly Nuclear Free Universe which would allow us to wage war on each other forever without the severe consequences of nuclear weaponry. Uranium238 and thorium are the only elements that can be bred into fissionable material, but they would decay too rapidly in a universe with the 1/3% more charged-up electrons.

Endnotes

Chapter 1

[1] *The Inflationary Universe*, Alan H. Guth, Addison-Wesley, N.Y., 1997.

[2] I have used a non-conventional spelling out of the isotopes followed by a number designating the number of protons and neutrons. I think this is much more user-friendly than the usual scientific abbreviation with the nucleon number as lower case. The idea is to communicate and not perplex the reader with unknown shorthand conventions.

[3] Resonant states are like the note produced by stringed musical instruments. The frequencies (alternately energy or temperature) have to be just right for the particles to resonate.

[4] The Snowball Earth hypothesis has recently obtained strong support from the analysis of rock strata in South Africa by J. Kirshvink. See *Science*, p 1342, 28 August 1998.

[5] *Rare Earth*, P.D. Ward and D. Brownlee, Copernicus, Springer-Verlag New York Inc., N.Y., 2000.

Chapter 2

[1] A. Sakarhov and S. Weinberg were the first to propose in the late 1960s that the surplus matter in the universe was connected to CP Asymmetry. More details are in an article by D.H. Freedman in *Science*, Vol. 254, p 3821, 1 October 1991.

[2] In 1963, Murray Gellman and George Zwieg proposed the Quark theory based on three different kinds of Quarks (Up, Down, and Strange). M. Kobavashi and J. Maskawa pointed out in 1973 that a theory with at least six kinds of quarks would allow for a CP violation.

[3] Many of these stepping-stones are discussed in *Cosmic Coincidences*, John Gibbin and Martin Rees, Bantam Books, N.Y., 1989.

Chapter 3

[1] *The Elegant Universe*, Brian Greene, Vintage Books, N.Y., 2000.

[2] *Three Roads to Quantum Gravity*, Lee Smolin, Basic Books, N.Y., 2001.

Chapter 4

[1] Four thousand different samples of fibroblast cells from endangered species are presently being stored in liquid nitrogen in the Frozen Zoo in the Center for Reproduction of Endangered Species at San Diego Zoo. Ultimately, the genetic viability of these endanger species can be enhanced by cloning new animals from this repository. Similarly, the diversity of life on Earth could be reestablished elsewhere using this technology.

[2] The Snowball Earth hypothesis has recently obtained strong support from the analysis of rock strata in South Africa. J. Kirshvink, *Science*, p 1342, 28 August 1998.

[3] Within a full range of 0.5% for the strong force or 4% for the eletromagnetic force, the production rate for carbon and oxygen vary by a factor of 30 to 1000. H. Oberhummer et al., *Science*, Vol. 289, p 88, 7 July 2000.

[4] *Cosmic Coincidences*, John Gibbin and Martin Rees, Bantam Books, p 244, N.Y., 1989.

[5] James Glanz, *New Data on 2 Doomsday Ideas, Big Rip vs. Big Crunch*, N.Y. Times, 21 February 2004.

Chapter 5

[1] Doris Lora, A Profile of Edgar Mitchell, *SHIFT*, Institute of Noetic Science, Pedaluma, Ca, p 19 , December 2003.

[2] James Lovelock and Lynn Margulis are prominent exponents of this way of thinking about the total ecosystem. Though Lovelock is a guru of the ecological movement, his support of nuclear power as a technically viable alternative to frying the planet is largely ignored. For more on Gaia see page 123 of *Rare Earth*, P.D. Ward and D. Brownlee, Copernicus, Springer-Verlag New York Inc., N.Y., 2000.

Chapter 6

[1] H. Ohmoto et al., *Science*, Vol. 262, p 555, 22 October 1993; *Science News*, p 276, 1 May 1935; J.S.R. Dunlop et al., *Nature*, Vol. 274, No. 5672, p 676, 17 August 1978; M.E. Barley et al., *Earth and Planetary Science Letters*, Vol. 43, No. 1, p 74, April 1979; and Walter et al., *Nature*, Vol. 284, No. 5755, p 443, 3 April 1980.

[2] D.W. Deamer and T. Smith, Biogenisis As An Evolutionary Process, *Journal of Molecular Evolution*, 33(3), p 207, 1991.

[3] This type of idea was first proposed by S. Arrhenius in 1903. He proposed the transport of bacterial spores (very tough dehydrated bacterial capsules) in meteors. So far none have been found.

[4] D. McKay et al., *Science*, Vol. 273, p 924, 16 Aug 1996.

[5] Cairn-Smith, A.J. Hall, M.J. Russell, Mineral Theories of the Origin of Life and an Iron Sulfide Example, *Origin Life*, 22(1–4), p 161, 1992.

[6] S. Rasmussen, et al., *Science*, Vol. 303, p 963–965, 13 February 2004.

[7] G.F. Joyce, Directed Molecular Evolution, *Scientific American*, 267(6), p 90, 1992; A.A. Beaudry, *Science*, Vol. 257, p 635, 1991; K.C. Nicolaou, *J. Am. Chem. Soc.*, Vol. 114, p 7555, 1992. Note that the earliest work on RNA replication with a protein was by Sol Spiegelman at the University of Illinois.

[8] D.P. Bartel, *Cell*, Vol. 67, p 529, 1991.

[9] Harold Morowitz (George Mason University, Fairfaz, Virgina) advocates this approach (also see[2]).

[10] Gunter Wachtershauser of Munich advocates this hypothesis. See Groundworks for an Evolutionary Biochemistry - The Iron Sulfur World, *Prog. Biophy*, 58(2), p 85, 1992.

[11] Kevin Zahnle of NASA Ames Research Center has proposed this. See: J.B. Pollach et al., Impact Generated Atmospheres Over Titan, Ganymede, and Callisto, *ICARIS*, 95(1), p 1, 1992; and Fractionization of Terrestrial Neon by Hydrodynamic Hydrogen Escape from Ancient Steam Atmospheres, *Meteorics*, 26(4), p 412 1991.

[12] M. Eiggen et al., *Sci. Amer.*, Vol. 244, No. 4, p 88, April 1981.

[13] C. Huber, et al., *Science*,Vol. 301, p 938, 15 August 2003.

[14] L. Margulis, *Symbiosis in Cell Evolution*, Freeman , San Francisco, 1981; S.J. Giovannoni et al., *J. Bacteriol*, Vol#170, 3584, 1988; and D. Yang et al., *Proc. Natl. Acad. Sci.*, USA, Vol#82, 4443, 1985.

[15] A.H. Knoll, "The Early Evolution of Eukaryotes: a Geological Perspective", *Science*, Vol. 256, p 622, 1 May 1, 1992.

Chapter 7

[1] J.R. Nursall, *Nature*, Vol. 183, p 1170, 1959; B. Runnegar, *J. Geol. Soc. Aust.*, Vol. 28, p 395, 1982; P. Cloud, *Paleobiology*, Vol. 2, p 351, 1976; A.H. Knoll, *Sci. Am.*, Vol. 265, p 64, October 1991.

[2] The term Snowball Earth was first used by Joseph Kirshvink of the California Institute of Technology in 1992 though many of the pieces of the concept had been previously proposed. Confirmation came from P. Hoffman et al., *Science*, p 1342, 28 August 1998. For a good recent summaries see R. Kerr, *Science*, p 1734, 10 March 2000.

[3] The creature called H. Bassetti has recently been found in Pennsylvania where it lived in swamps 365 million years ago. E.B. Daeschler, Science, Vol. 265, 8 July 1994.

[4] A.H. Knoll et al., Comparative Earth History and Late Permian Mass Extinction, *Science*, Vol. 273, 26 July 1996.

[5] The Siberian traps was first proposed as the source of the Permian extinction by Paul R. Renne of the Institute of Human Origins in Berkeley California and Asish R. Basu of the University of Rochester in 1991. Accurate dating was reported by Gerald K. Czmanske in *Science*, December 1992.

[6] A. Basu et al., *Science*, Vol. 302, p 1388 (2003).

[7] Evidence of an excess of Iridium in a clay layer that marks this event was reported by Walter and Loui Alverez, F. Asaro, and H.V. Michel, *Science*, Vol. 208, p 1095, 1980. Additional supporting evidence came from C.C. Swisher et al., *Science*, Vol. 257, p 954, 1992. Evidence for the crater was reported by V.L. Sharpton et al., *Science*, Vol. 261, p 1564, 1993.

[8] H.J. Melosh (University of Arizona) has proposed that the impact generated nitric acid that caused a severe case of global acid rain. O.B. Toon (NASA Ames Research Center) has suggested that global forest fires might be the cause. Kevin Pope (Geo Eco

Arc Research in La Canada Calif.) hypothesized that sulfuric acid clouds could have caused a ten year winter.

Chapter 8

[1] G. Vogel, *Science*, Vol. 303, p 1128, 20 February 2004.

[2] *The Phenomenon of Man*, Pierre Teilhard de Chardin, Harper and Row, N.Y., 1965.

[3] For a recent review of whale and dolphin vocalizations and culture see Bruce Bower, *Science News*, Vol. 158, 10/28/00, p 284.

[4] Illustration by Caroline Mallary, 2003.

[5] R. Dawkins has coined the word "meme" to label simple mental entities of this nature in *The Blind Watchmaker*, Penguin and Norton, N.Y. and London, 1987. The concept goes back many decades and has many names. Also see S. Blackmore, *Scientific American*, p 66, October 2002.

Chapter 9

[1] *Echo of the Elephants*, Cynthia Moss and Martyn Colbeck, William Morrow and Company, New York, 1992.

[2] Koko was trained by Francene Patterson of the Gorilla Institute. Koko swears, gossips, jokes, and lies.

[3] Katherine Payne of Cornell University Laboratory of Ornithology discovered elephant infra-sounds after first feeling them at a zoo in Portland.

[4] Ibid [1], *Echo of the Elephants*, p 41.

[5] Ibid [1], *Echo of the Elephants*, p 60.

[6] Ibid [1], *Echo of the Elephants*, p 94.

[7] Elephants sometimes have a problem with alcohol. They occasionally bury fruit and eat it after it has fermented. Rouge elephants, marauding a village, are often drunk.

Chapter 10

[1] *National Geographic*, p 593, May 1978.

[2] The tribal people of the Andaman Islands in the Indian Ocean have been isolated for 50,000 years. They use fire but they do not know how to start it. *Scientific American*, p 16, May 1995.

[3] R.L. Cann, et al., *Nature*, Vol. 325, #6099, p 31, 1 January 1987. L. Tianyuan and D. Etler, *Nature*, 4 June 1992. J. Klein, N. Takahata, F.J. Ayala *Scientific American*, p 78, December 1993.

[4] *The Age of Mammals*, Bjorn Kurten, Columbia University Press, N.Y., 1972.

[5] *Science News*, p 91, 8 February 1992.

[6] G. Dehaene et al., *Nature*, 28 July 1994.

[7] The Aztecs solved this one by taking live prisoners back to be eaten by their nobles.

[8] Polygamy in their society creates a high demand for brides. Potential brides are often kidnapped when a conniving father promises his daughter to more than one suitor. The cycles of revenge that results make being a young man a hazardous lifestyle. See *Yanomamo, the Fierce People*, Napoleon Chagnon, Holt-Rrinhart-Winston, N.Y., 1977. Note that there is controversy over whether or not warfare amongst these people was in response to contact with outsiders such as Chagnon.

[9] V. Morell, *Science*, Vol. 265, p 312, 15 July 1994.

Chapter 11

[1] Faye Flam, *Science*, Vol. 265, p. 1032, 19 August 1994.

[2] *The Blind Watchmaker*, R. Dawkins, Penguin and Norton, N.Y. and London, 1987. Also see S. Blackmore in *Scientific American*, p 66, October 2002.

[3] *The Minds Sky*, Timothy Ferris, Bantam Books, N.Y., 1992.

[4] Thomas S. Ray, University of Delaware, created the program called Tierra. It is a cybernetic ecosystem. It was reported in *Scientific American*, p 132, January 1992.

[5] M. Nowak et al., The Arithmetics of Mutual Help, *Scientific American*, Vol. 272, #6, p 76, June 1995.

[6] M. Nowak and K. Sigmund, *Nature*, 1July1994. and *Science News* Vol. 144, p 6, 3 July 1994 and *Science News*, p 39, 18 January 1992.

[7] M. Milinski, *Nature*, 1 July 1994.

Chapter 12

[1] A. Lazcano in *Science*, Vol. 299, p 347, 17 January 2003. Also see *Life Evolving –Molecules, Mind and Meaning*, C. de Duve, Oxford University Press, New York and London, 2002.

[2] *Science*, Vol. 287, p 1743, 10 March 2000.

[3] Cosmology, Edward Harrison, Cambridge University Press, Cambridge, p 399, 1981.

Chapter 13

[1] Private communication with B. Buffett, Dept. of Geophysical Sciences, U. of Chicago. Also see A. Buffett, *Science*, Vol. 299, p 1675, 14 March 2003. At this time, the proportion of heat from uranium, thorium, potassium, and residual impact energy are estimated to be about 20, 20, 35, and 25%, respectively.

[2] *Principles of Modern Physics*, Robert Leighton, McGraw-Hill, N.Y., 1959. Application of Eq(15.9) on p 527 yields the functional dependence of alpha decay rate on the electron charge for heavy nuclei (see Appendix I).

Chapter 14

[1] A. Linde, *Scientific American*, p 48, November 1994.
[2] Nuclear holocaust was a lot closer than anyone had realized. Unknown to the American Joint Chiefs of Staff, who advocated invasion, the USSR had four times as many troops in Cuba as the intelligence community reported, and they were armed with tactical nuclear weapons. It is likely that an invasion would have resulted in a tactical nuclear war in which tens of thousands of Americans would have died at the outset. Escalation would have been very likely if President Kennedy had taken the advice of the Joint Chiefs.

Bibliography

The Anthropic Cosmological Principle. John Barrow and Frank Tipler. Oxford University Press, N.Y. and London, 1986.

The Ascent of Man, British Broadcasting Company, J. Bronowski, London, U.K. 1973.

The Age of Mammals, Bjorn Kurten, Columbia University Press, N.Y., 1972.

Beyond Einstein, The Cosmic Quest for the Theory of the Universe, M. Kaku, and J. Thompson Anchor Book, Division of Random House, N.Y., 1995.

The Blind Watchmaker, R. Dawkins, Penguin and Norton, N.Y. and London, 1987.

A Brief History of Time, Stephen Hawking, Bantam, N.Y. and London, 1988.

Coming Of Age In The Milky Way Galaxy, Timothy Ferris, William Morrow Company, N.Y., 1988.

The Creation, P.W. Atkins, W.H. Freeman and Company, San Francisco, Calif., 1981.

Cosmic Coincidences, John Gibbin and Martin Rees, Bantam Books, N.Y., 1989.

Cosmos, Carl Sagan, Random House, Inc., N.Y., 1981.

Dreams of a Final Theory, Steven Weinberg, Pantheon Books – Random House, Inc., N.Y., 1992.

Echo of the Elephants, Cynthia Moss and Martyn Colbeck, William Morrow and Company, N.Y., 1992.

The Elegant Universe, Brian Greene, Vintage Books, N.Y., 2000.

The First Three Minutes, Steven Weinberg, Bantam New Age Books, N.Y., 1977.

Galaxies, Nuclei, and Quasars, Fred Hoyle, Harper and Row, N.Y., 1965.

The Global Brain, Peter Russell, J.P. Tarcher, Inc., Los Angeles, 1983.

God and the Astronomers, Robert Jastrow, Wagner Books, N.Y., 1978.

God and the New Physics, Paul Davies, Simon and Schuster, N.Y., 1983.

The God Particle, Leon Lederman, Hought and Mifflin, N.Y., 1994.

Infinite in All Directions, Freeman Dyson, Harper and Row, N.Y., 1988.

The Inflationary Universe, Alan H. Guth, Addison-Wesley, N.Y., 1997.

The Intelligent Universe, Fred Hoyle, Michael Joseph, London, 1983.

Life Evolving–Molecules, Mind and Meaning, C. de Duve, Oxford University Press, New York and London, 2002.

Messages from the Stars, Ian Ridpath, Harper and Row Publishers, N.Y., Hagerstown, San Francisco, London, 1978.

The Mind of God, Paul Davies, Simon and Schuster, N.Y. and London, 1992.

The Minds Sky, Timothy Ferris, Bantam Books, N.Y., 1992.

The Moment of Creation, James S. Trefil, Charles Scribner's Sons, N.Y., 1983.

The Origin Of The Universe, John Barrow, BasicBooks- HarperCollins Publishers, N.Y., 1994.

The Phenomenon of Man, Pierre Teilhard de Chardin, Harper and Row, N.Y., 1965.

Principles of Modern Physics, Robert Leighton, McGraw-Hill, N.Y., 1959.

Rare Earth , P. D. Ward and D. Brownlee, Copernicus, Springer-Verlag New York Inc., N.Y., 2000.

The Seven Mysteries of Life, Guy Murchie, Houghton Mifflin Comp., Boston, Ma, 1978.

The Symbiotic Universe, George Greenstein, William Morrow, N.Y., 1988.

Three Roads to Quantum Gravity, Lee Smolin, Basic Books, N.Y., 2001.

What is Life?, Erwin Schrodinger, Cambridge, New York, Melbourne, 1992.